CLIMATE CHANGE

Our Warming World

Christina Hutchins

Introduction by
Archbishop Desmond Tutu

Author: Christina Hutchins
Picture Selection: Christina Hutchins
Project Manager: Lyn Hemming
Design: Tom Germain
Production: Julia Richardson, Anny Mortado
Editor: Katherine Pate

First edition of Climate Change – Our Warming World

Copyright © 2009 Alastair Sawday Publishing Co. Ltd.
Text copyright © Christina Hutchins
Fragile Earth an imprint of Alastair Sawday Publishing Co. Ltd.

First published in November 2009

Alastair Sawday Publishing Co. Ltd
The Old Farmyard, Yanley Lane,
Long Ashton, Bristol BS41 9LR
Tel: +44 (0)1275 395430
Fax: +44 (0)1275 393388
Email: info@sawdays.co.uk or info@fragile-earth.com
Web: www.sawdays.co.uk or www.fragile-earth.com

ISBN-13: 978-1-906136-31-4
Printed in United Kingdom by Butler Tanner & Dennis, Frome
UK distribution: Penguin UK, London

For Maya and the children of her generation

"

Scientific evidence on various aspects of climate change is now very strong, highlighting the need for urgent action not only for mitigating the emissions of greenhouse gases but also for implementing measures involving adaptation to the impacts of climate change. These actions would only be successful if communities and people at large were to take action on their own, irrespective of whatever governments and international bodies might do. This book raises awareness about climate change and urges crucial actions that need to be taken at the level of communities, families and individuals to prevent catastrophic changes to our planet.

Dr Rajendra K. Pachauri
Chairman, United Nations Intergovernmental Panel on Climate Change

PREFACE

I built my first windmill in 1996. How could I forget? Not only was it the end of a five-year struggle but it went up on Friday 13 December. A year later I went to the UN conference in Kyoto, the one that put climate change on the map. I turned up with a friend with no idea what to expect. Our plan was to talk anyone that would listen about the role wind energy has to play in fighting climate change. The message went down well.

Back then hardly anyone had heard of climate change and even fewer people held it to be true. It really was the territory of scientists and 'hippies'. Actually, almost 2,000 of the world's top scientists had told our political leaders "There is only one responsible choice – to act now." It was an exciting time. And Kyoto was a success. Not because the targets agreed were particularly ambitious but because targets were set at all. It was and still is an important moment in human history. The world had set its first carbon targets.

Much has changed since then. Most people today have heard of climate change and more importantly accept that we urgently need to do something about it. It's as rarely out of the news these days as it was rarely in the news a decade ago. The Kyoto accord played a vital part in this transformation: many people are hoping that a successor to Kyoto will emerge from Copenhagen in December 2009.

But Kyoto targets have not been met. What use are another set of targets? We actually need deeds now, not words. While we've been busy talking about climate change, it's been busy demonstrating what it's capable of.

I write this in July 2009. Outside, hailstones the size of peas are bouncing off the pavements. This is the UK's third 'monsoon July' in a row, and nobody's idea of summer. In truth though, the west is experiencing relatively little of climate change – in other parts of the world it's destroying lives. People far away are living and dying with its worst effects. This book is here to bring that truth into our living rooms. It's an important truth. The choices we make are having a profound impact. We can't see it so clearly in our lives, but we can see it, through this book, in the lives of others.

80% of our personal carbon footprint comes from choices we make, the things we choose to spend our money on – the things we can change. We can all make significant reductions with simple changes – to the way we power our homes, how we travel and what we eat.

This book is about the consequences of not acting; it's about a harsh reality that our choices are forcing on other people. The imperative to act is clear. To fail to act, when the means are within our easy grasp, would be unconscionable in my opinion.

Would I be able to look into the eyes of the woman on page 131 and explain why I used the car for a journey of under a mile today? Could I tell the child on page 51 why I buy food flown from the other side of the world? Would the man on page 107 say "Hey – don't worry about it," if I told him I've been too busy to get around to switching my electricity supply to wind-power? Unlike them, we have the luxury and the power of choice. Let's use it.

Dale Vince
Founder of Ecotricity

PS Switching to wind power is easy – see the panel on the inside back cover of this book for details. Sign up to Ecotricity and we'll even give you back the cover price of this book.

Photo: Ecotricity

CONTENTS

INTRODUCTION

Climate change is the greatest challenge we face today – in the coming decades the world as we know it may be changed. Already we see the impacts with more floods, droughts and species threatened. And Africa is to be one of the regions most badly affected – certain areas, such as southern Africa, are projected to become more and more drought stricken, which will worsen the severe famine and drought of recent years.

Throughout history there have been many times when people came together to call for change – Gandhi and his fight for India and in my homeland our struggle against apartheid. So too today humankind faces a challenge on an

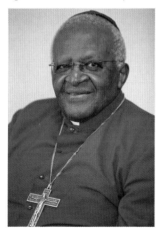

unprecedented scale – within the coming years we have the choice either to watch our planet be irrevocably altered or to come together internationally and strive for change. What is the price if we do not do so? Hundreds of millions of people going hungry, many of these people in Africa; billions suffering a lack of water – the very basis of life; species across the globe being forced into extinction. Is this the world we seek?

I ask those that read this book to do all they can to bring about change. To do what they can individually and to come together collectively to protect the future of our world, for our children, for the sake of the most vulnerable and for the generations to come.

Archbishop Desmond Tutu

Our climate is changing and humans are responsible.

OUR WARMING WORLD

The world's leading scientists agree that human actions are causing the Earth to get warmer. Through our use of oil, coal and gas to supply the world's energy we are increasing the heat-trapping gases in our atmosphere so raising our planet's temperature. This in turn is changing our climate and threatens to change our world.

The average temperature of the Earth has only risen by about 0.8°C since 1900, but already heavy rain and droughts are increasing, ice sheets melting, sea levels rising and hurricanes are more severe in some regions. More than 300,000 people die annually due to climate change causes and an estimated 325 million are seriously affected every year.

By 2100 the world could warm by between 1.6° and 6.9°C above 1850–1899's global average. With a 2° to 4.6°C rise polar ice sheets would melt significantly, seas would inundate coastal settlements and floods and droughts could dramatically increase. Water resources and food production are likely to decline in many regions and up to 50% of species could face extinction. All of humanity will be affected, including our children.

There is still a small window of time to prevent this. Global emissions of carbon dioxide must stop rising or 'peak' by 2015 or before if the dangerous warming of the Earth is to be prevented. The solutions exist to achieve this, but government actions fall far short. Only if the world's people each take action as a global movement may change occur in time.

The Earth is in our hands today and your help is very much needed.

Photo: Anders Pettersson – UNEP/Still Pictures

Life on Earth is thought to have begun with a solitary cell deep in the oceans over 3.5 billion years ago. And from that first cell life has evolved on this planet.

From the deepest of oceans to the richest of rainforests, from the smallest bacteria to the largest of mammals, our world is beautiful and pulsates with life.

And yet now within one human lifetime we are on course to raise the world's temperature to such a high level and so rapidly, that it threatens our planet as we know it.

Photo: Frans Lanting/Corbis

A changing atmosphere

It is our atmosphere that sustains us. A thin blanket of air around the world on which we all depend – a mix of gases so finely balanced they enabled life to flourish on Earth. By trapping just enough of the sun's heat, like a greenhouse, our atmosphere keeps our planet 30°C warmer than it would otherwise be. Without this natural greenhouse effect the Earth would not be habitable. Mars with its thin atmosphere, comprising mostly carbon dioxide, is a freezing −63°C, while Venus's atmosphere, over 50 times more dense than Earth's at the surface and also mostly carbon dioxide, is a boiling 464°C.

The mix of natural greenhouse gases that makes up Earth's atmosphere has remained largely stable for the last 10,000 years or so and has been critical in providing and sustaining our life-supporting temperatures and climate. But since the beginning of the industrial era in 1750, human activities have released billions of tonnes of additional greenhouse gases. And every hour of every day, everywhere, we are pumping more and more of these gases into our fragile atmosphere. The delicate balance of gases is being altered and the new mix traps more heat, so raising the temperature of our planet.

Photo: Yanik Chauvin/iStockphoto.com

Rising temperatures

The ten hottest years since records began in 1850 have all occurred since 1996. And all of the 20 warmest years have occurred since 1980. Globally, 1998 and 2005 were the hottest years on record – followed by 2003, 2002 and 2004.

The average temperature of the world has risen by about 0.8°C since 1900. But this average conceals the fact that many regions have warmed much more than this. Since 1970 northern Africa, Spain and parts of Russia and China have seen temperatures rise by 1° to 2°C. Canada and Alaska are now 3° to 4°C warmer in winter than in the 1950s – that's four to five times more than the global average temperature increase. So if the world's average temperature rises by 3°C, Canada and Alaska could heat up many times more.

Already some countries are getting too hot for human existence. In May and June of 2005 parts of India and Pakistan saw temperatures rise to 50°C, causing hundreds of deaths. By 2100 these regions could be 2° to 5°C hotter. This suggests that in time whole regions of the Earth could become too hot for human existence.

As the world warms, forest fires are increasing in some regions.

The speed of change

What is of such concern today is the rate at which the average temperature of the world is rising and how this rate is accelerating. It is unprecedented in our record of past temperatures on Earth.

Ice core records reveal a repeated pattern of cold glacial periods followed by warmer interglacial periods, with each glacial–interglacial cycle lasting about 100,000 years. Between ice ages and warm interglacial periods the Earth's average temperature rose 4° to 7°C, but this took about 5,000 years to happen. Today the average temperature of the world is rising much faster – by about 0.2°C every decade. In just over one human lifetime, by 2100, the Earth´s average temperature could rise another 1.1° to 6.4°C. This would be additional to the warming to date, creating a potential increase of up to 7°C since 1900. It could match the highest temperature rise between the glacial–interglacial periods, but this time in just 200 years.

Temperature changes in certain regions are likely to be even greater. Average Arctic temperatures have increased at almost twice the global average rate in the past 100 years and land areas are warming twice much as the oceans.

We are turning up the world's thermostat but won't be able to turn it back down.

Photo: REUTERS/Stringer India

The key cause

Human-induced global warming is caused by billions of tonnes of greenhouse gases released into Earth's atmosphere due to human activities. Carbon dioxide is the most important greenhouse gas, released primarily by burning oil, coal and gas to supply the world's energy, including electricity and transport.

Oil, coal and gas are fossil fuels – the fossilised remains of dead trees, plants and animals that died up to 300 million years ago. Like a global pyre we burn these remains to provide energy for humankind, releasing their stored carbon. Fossil fuels and cement production account for about 80% of the carbon dioxide released by human activities. The other 20% comes from clearing forests and vegetation – as fallen plants decompose or burn, carbon dioxide is released into the atmosphere.

Methane and nitrous oxide are also greenhouse gases. Sources include industrial processes, decaying rubbish in landfill sites, livestock belching, biomass burning and nitrogen-based fertilisers. All contribute to human-induced global warming. Additionally, as the world warms due to human activities, water vapour increases in the atmosphere. This natural greenhouse gas intensifies the greenhouse effect further, causing more warming and more water vapour in a self-reinforcing cycle.

However, since carbon dioxide is the most important greenhouse gas emitted by human activities, this book focuses on this greenhouse gas.

Drax Power Station in North Yorkshire, England provides about 7% of the UK's electricity.

Photo: Jason Hawkes

Carbon emissions

On Earth, carbon is everywhere we look and is one of the basic building blocks of life. About 760 billion tonnes are combined with oxygen as carbon dioxide in our atmosphere; over 2,000 billion tonnes are stored in trees, vegetation and soils and almost 40,000 billion tonnes are stored in the world's oceans. Carbon is constantly exchanged between the land and atmosphere and between the oceans and atmosphere, moving each way every year in a natural carbon cycle.

However humankind is now disrupting the atmosphere's carbon content by emitting more and more carbon. In the 1980s human activities released an average of about seven billion tonnes of carbon annually. By the 1990s this had risen to about eight billion tonnes, and between 2000 and 2005 it rose further, to an average of about nine billion tonnes each year. Fortunately not all of this carbon has remained in the atmosphere. About half has been absorbed by the oceans, trees, vegetation and soils, so called 'carbon sinks', which have prevented the world warming still faster.

As a result an average of about 3.3 and 3.2 billion tonnes of extra carbon was left in the atmosphere each year in the 1980s and 1990s respectively, increasing the overall total. Between 2000 and 2005 this increased by over 30% to an average of about 4.1 billion tonnes per year.

   Photo: Barry Lewis/Corbis

Diminishing carbon sinks

As the world warms more and more, the oceans, trees, vegetation and soils, which currently absorb about half of the carbon dioxide produced by human activities, are likely to absorb less and less. This will leave more carbon dioxide in the atmosphere, increasing the world's average temperature by 0.1°C to 1.5°C. Still worse, as temperatures continue to rise, forests and soils are expected to release their own stores of carbon through increased respiration, raising temperatures still higher. In time some forests may die back, causing more carbon dioxide to enter the atmosphere as dead trees decay. The release of additional greenhouse gases from other natural sources such as permafrost, peatlands, wetlands and ocean stores of methane could raise temperatures even more.

And this could all occur just as human activities release increasing amounts of carbon dioxide. It would mean that the amount of carbon dioxide in the atmosphere could dramatically increase, causing the Earth to warm even faster than it is doing today. Already since 1995 the rate at which the carbon dioxide content of Earth's atmosphere is rising has increased by 35%. So the problem is rapidly getting worse.

Photo: Phil Schermeister/Still Pictures

The crisis we face

Tiny bubbles of air trapped in ice up to 3km deep at the South Pole reveal how temperatures have closely correlated with the carbon dioxide content of the atmosphere over 800,000 years. These ice core records show that between ice ages and warm interglacial periods, Antarctic temperatures rose and fell by 12°C. At the same time the amount of carbon dioxide in the atmosphere similarly rose and fell by about 100ppm (100 molecules of carbon dioxide for every million molecules of air), ranging from 180ppm to about 280ppm. Temperatures are believed to have changed first, with carbon dioxide levels following a few hundred years later.

Today carbon dioxide is increasing first, with temperatures following. Since the beginning of the industrial era the amount of carbon dioxide in our atmosphere has risen over 100ppm, from about 280ppm to 387ppm. This is increasing by almost 2ppm per year and by 2100 could, in the worst case scenario with high emissions, rise to 1200ppm as the graph shows. This would be over 900ppm above the pre-industrial level of 280ppm and would constitute over nine times the total increase of 100ppm of the last 800,000 years. Just that 100ppm caused a 12°C change in Antarctic temperatures.

This graph shows temperature (blue line) and carbon dioxide (CO_2) concentrations in the atmosphere (red line) over the past 420,000 years. The correlation between the two extends back 800,000 years before the present. By 2100 the amount of CO_2 in Earth's atmosphere could be (A) 1200ppm with continued intensive fossil fuel use or (B) 350ppm to 400ppm if emergency globally co-ordinated policy action is taken now to arrest climate change.

Photo: Petit, J.R. et al (1999) 'Climate and Atmospheric History of the past 420,000 years from the Vostok Ice Core Antarctica' *Nature* 399: 429-436

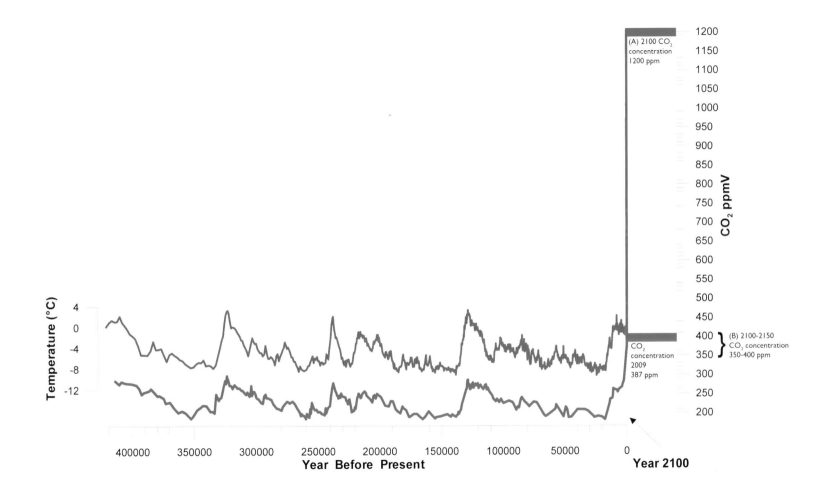

Surprises

The United Nation's Intergovernmental Panel on Climate Change, comprising the world's leading scientists, is assessing how our climate and world may change as carbon dioxide increases in the atmosphere, causing the Earth to warm. Their reports, released every five to seven years, reveal that much is now known about the climate system. At the same time, there could be surprises:

Future unexpected large and rapid climate system changes (as have occurred in the past) are, by their very nature, difficult to predict. This implies that future climate changes may also involve 'surprises'.

Intergovernmental Panel on Climate Change

One such 'surprise' could be an abrupt and rapid temperature rise over Greenland, which happened repeatedly over the North Atlantic region during the last glacial period, when Greenland warmed by up to 16°C in just a few decades before cooling again. This shows what is possible. Parts of Greenland have already warmed by 2.5°C, three times the world's average temperature increase of 0.8°C.

So if the world's average temperature rises by 3° to 6°C this century, temperatures over Greenland could warm considerably more, causing rapid ice melt, sea level rise and disruption of the major ocean currents that transfer heat from the tropics to polar regions.

Climate sensitivity

Another climate change 'surprise' could involve the Earth's response to increasing levels of carbon dioxide. Scientists affirm that if the amount of carbon dioxide in the world's atmosphere doubles from 280ppm to about 550ppm, as expected by 2050 if we do not act, the average temperature of the world will most likely increase by 3°C. The likely range is 2° to 4.5°C and it is very unlikely to be less than 1.5°C. At the same time values considerably higher than 4.5°C cannot be ruled out, although models agree less for higher values. Even a 3°C rise would cause global disaster – and higher than this would be cataclysmic.

And even when humans stop emitting greenhouse gases, the Earth will continue warming and sea levels will continue rising for more than a thousand years. This is because only 80% of the carbon dioxide we have added to the atmosphere will be removed in a few centuries: about 20% is likely to stay there for thousands of years, sustaining increased warming. This means we must act now if we are to avoid committing future generations to catastrophe.

A man tries to make his way through a lake of dead fish caused by severe drought in Brazil's Amazonia region (2005).

The time lag

The effects of climate change that we are witnessing today are the result of greenhouse gas emissions from 30 to 50 years ago. Since the mid-1970s greenhouse gas emissions and the rate of warming have both increased, meaning that we are already committed to much greater and more severe climate impacts in the decades to come.

This time lag means that the emissions we choose to release or limit now will profoundly affect the climate from 2050 and beyond. Additionally, since 20% of the carbon dioxide we add to the atmosphere is likely to stay there for thousands of years, our emissions today will continue to affect the Earth's climate for millennia.

Because of this time lag, our generation will not see the worst consequences of our emissions, but our children and grandchildren will be forced to live with them. So we must act now if we seek to help them.

In November 2004, an estimated 100 million pink locusts swarmed the Spanish Canary Island of Fuerteventura. As temperatures rise, the distribution of pests may change.

Photo: REUTERS/Juan Medina

A CHANGING CLIMATE AND CHANGING WORLD

Climate can be described as long-term average weather – the average and variability of temperature, rainfall and wind over a length of time, ranging from months to millions of years but classically specified over 30 years. Climate and extreme weather events are determined by the close and complex interaction between the atmosphere, the land, oceans, snow, ice and living things, so that a change in one system affects the others.

Changes in climate and extreme weather events occur all the time, making it difficult to attribute any one severe weather event to human-caused global warming. But human activities are now considered responsible for most of the warming of the past 50 years. And despite natural variability some weather events, such as heavy rainfall and heatwaves, have become more frequent and intense as the world has warmed. Europe's heatwave of 2003 lay so far outside the range of natural variability that human influence is believed to have contributed to its severity.

As our climate changes it is affecting the natural world, causing ice melt, sea level rise and ecosystem loss. As temperatures continue to increase, chaotic and unpredictable impacts could cascade across the globe, causing potentially rapid and irreversible changes.

Photo: Jim Reed/Corbis

During the 20th century, precipitation increased by 10% to 40% in parts of northern Europe and decreased by up to 20% in some regions of southern Europe.

Torrential rain sent a 2.5 metre wave surging through the village of Boscastle, England (August 2004). The flash flood swept cars out to sea, buildings collapsed and residents were left stranded and had to be air-lifted to safety. River levels rose 2 metres in just 60 minutes.

Precipitation

As the world warms, substantially heavier rain is falling more frequently over many land areas.

This is because for every 1°C rise in temperature the atmosphere can hold about 7% more water, and also warmer oceans and land surfaces cause more powerful evaporation – more moisture ascends into the atmosphere. The combined effect is a more intense water cycle, with a greater proportion of total precipitation falling in heavy and very heavy downpours.

Central and eastern North America and northern and central Asia have seen precipitation increase by 6% to 8% since 1900, while in the Northern Territory of Australia annual total rainfall increased by 18% from 1910 to 1995. Northwest China saw rainfall increase by 22% to 33% between 1961 and 2000, while in the United States very heavy precipitation has increased by 20% with much of the increase occurring in the last three decades of the 20th century. In India storms releasing 15cm of rain in a day have doubled in number since 1950.

By 2050, annual average river runoff is expected to increase by 10% to 40% in northern regions of the world and in some wet tropical areas. River runoff is all the water that accumulates from rainfall and melt-water, which feeds into the seas. And as temperatures rise, for every 1°C the world warms, the intensity and amount of rainfall can be expected to increase.

Photo: REUTERS/Romeo Ranoco

England and Wales 2007

In 2007 England and Wales experienced the wettest May, June and July since records began in 1766. Rainfall was double the seasonal average, inundating tens of thousands of homes and thousands of businesses. 13 people died and one million people were affected. 17,000 homes were evacuated and up to 50,000 homes were without electricity. 420,000 properties were without drinking water, some for up to two weeks. By the end of May 2008 local authorities estimated that almost 5,000 people had still not been able to return to their homes. The floods caused £3 billion in damages.

Yet with high carbon dioxide emissions winter precipitation in some parts of the UK could increase by 30% to 50% towards the end of this century.

England (July 2007). The town of Tewkesbury surrounded by flood waters, following torrential rain.

Photo: Daniel Berehulak/Getty Images

Testimonial: Tewkesbury Floods, 2007

Margaret Martin, Longlevens near Tewkesbury 8 miles from Gloucester

In my whole life I have never seen as much rain as that which fell this summer. Our house is 51 years old and has never flooded before. On that Friday morning a big puddle formed in the road and kept getting bigger. By 1 o´clock a lot of water came gushing into my garden and then water started coming out from under the cooker. I burst into tears. I kept running back and forth with my husband, taking as much as possible upstairs – the computer, TV, chairs, etc. By 8 o'clock the water was quite deep in the house. We decided to leave. As we waded out through the front door the water was up past my knees with about 20 houses flooded – including my 80-year-old neighbour's house.

The flood waters remained till the Sunday leaving a damp nasty smell from the sewage that had been in my house. We had no electricity, no hot water, no gas and the water was cut off for nine days. We lived upstairs in these conditions as the house dried out, was re-wired and plastered and a new kitchen put in. Water was provided in big containers on the street or bottles from the supermarket. Some people had to live in caravans. People were very demoralised as everything took weeks to do. And now every time it rains I feel nervous, thinking it may happen again. I also think about my grandchildren and wonder: What are we going to do?

Photo: Gideon Mendel/Corbis

The proportion of the Earth's total land area suffering extreme drought is expected to increase dramatically from 1% in the 20th century to up to 30% by 2100 with continued high emissions.

Drought

As the water cycle becomes more intense, the distribution of rainfall is changing: more rain is falling in some areas and less in others. Higher temperatures also cause land areas to dry out more. The combined effect is that dry areas across the world have more than doubled since the 1970s, with more intense and longer droughts occurring over larger areas, particularly in the tropics and sub-tropics.

By 2050 annual average river runoff and water availability could decrease by 10% to 30% in some dry land areas and in the dry tropics – some of which are already suffering water shortages. Southern and eastern Australia, northern Africa and Latin America, southwest North America and many small islands in the Caribbean and Pacific are all likely to have less rain. In certain regions of the UK summer precipitation could decrease by up to 60% with continued high emissions and up to 70% in southern and central Europe.

In northern and eastern Africa and North America climate records show that in the past, abrupt changes in rainfall patterns have triggered droughts lasting decades or even hundreds of years. Already the southeast of the US is suffering one of the most severe droughts ever recorded in that region – a possible precursor of worse still to come.

Photo: Bob Thomas Bettmann/Corbis

Africa

Already about 200 million African people annually experience high water stress, defined as less than 1,000m^3 of water available per person per year. Climate change is expected to worsen the water crisis in many African countries and cause water stress in others that currently do not experience it. By mid-century 350 to 600 million African people could suffer increased water shortages.

Rainfall has already decreased in some regions. West Africa has seen a 20% to 40% reduction since the end of the 1960s and southern Africa's dry season is now longer, with less predictable rainfall. As a result, 14 million people have suffered severe drought and famine in recent years. By 2050, southern Africa could have up to 30% less water, and 50% less if the Earth's average temperature rises by 4°C. Northern Africa could have up to 75% less water with this degree of warming.

By as early as 2020, climate change could cause increased water stress for between 75 and 250 million African people and in some African countries yields of rain-fed crops could decline by up to 50%. This will exacerbate hunger, malnutrition and death. Additionally Africa's population is expected to double to almost two billion by 2050 – a billion extra people with much less food and water than now.

Photo: Jon Jones/Sygma/Corbis

By 2050 European heatwaves could be 4° to 10°C hotter and more frequent.

 Photo: REUTERS/Albert Gea

Heatwaves

In the last 50 years, cold days, cold nights and frosts have become less frequent, while hot days, hot nights and heatwaves have increased across the world. As the Earth warms this trend is likely to continue, causing fewer cold-related deaths but more heat-related fatalities as heatwaves increase in frequency, duration and severity.

In the US heat-related deaths could more than double by 2020. In 2006 Los Angeles recorded its highest temperature of 48°C, with up to 54°C in other parts of California. By 2100 rising temperatures could cause heat-related mortality in Los Angeles to increase seven-fold and with a 3°C increase there could be a net 86,000 extra deaths a year across Europe.

In the UK, twice as many hot summer days now exceed 25°C than during the first half of the 20th century, resulting in more heatwaves. The most notable was the European heatwave of 2003, the hottest summer since records began, with temperatures 6°C above the long-term average hitting 40°C in some regions. Over 45,000 people died from the heat, the majority of them elderly. By 2040 this hot a summer is expected to be normal and by 2060 will be considered cool.

Photo: Spencer Platt/Getty Images

The power of hurricanes in the North Atlantic and western North Pacific has almost doubled since the mid-1970s.

A satellite image shows Hurricane Katrina approaching the Louisiana and Mississippi Gulf Coast.

Hurricanes

As temperatures rise, global wind patterns and the strength of winds are being affected. In the southwest Pacific Ocean the most powerful hurricanes – category 4 and 5 – have more than doubled in number since the 1970s. In other regions the power released by hurricanes has increased, partly due to warmer sea surface temperatures (SSTs) fuelling the storms' energy. Hurricanes and typhoons only develop when SSTs exceed about 26°C, so a warmer world is likely to extend the range of such storms and increase their intensity.

In 2005, high SSTs in the North Atlantic Ocean coincided with the record North Atlantic hurricane season, producing double the annual average number of storms since 1995 and triple the average before that, including the first hurricane to make landfall in Spain and Portugal. Three of the ten strongest hurricanes ever were recorded, including the most powerful – Hurricane Wilma – and the most expensive – Hurricane Katrina, with US$135 billion in economic losses. New Orleans was 80% flooded with sea water up to 6 metres deep, one million inhabitants were evacuated and over 1,800 people died.

Storms causing US$100 billion of insured and US$200 billion of economic losses are now envisaged. This would cripple key sectors of the global economy and put many more millions of lives at risk.

In 2007 nearly 50% of the Arctic's normal summer sea ice cover melted – an area four times the size of France and Germany combined.

Photo: Norbert Rosing/National Geographic Stock

The world's ice cover

The Earth's ice cover, the cryosphere, forms the second largest component of the global climate system after the oceans. The presence or loss of ice and snow in polar regions is associated with changes in temperature, which affect winds and ocean currents. Ice permanently covers 10% of the Earth's land surface and about 7% of the oceans as an annual average, while land ice stores about 75% of the world's fresh water.

The Earth's ice cover is providing the clearest evidence of climate change – it is shrinking worldwide. Glaciers and ice caps are receding, ice shelves collapsing, snow cover decreasing, sea ice melting and by 2020 Mount Kilimanjaro's ice cap could disappear for the first time in 11,000 years.

And as the ice on our planet melts, light white surfaces of ice and snow, which reflect 90% of the sun's rays back into space, are replaced by darker ocean or land surfaces, which reflect only 10%. As more dark surfaces are exposed as ice melts, the Earth absorbs more of the sun's heat, increasing the warming and melting and so on, in a dangerous cycle that accelerates global warming.

Melting glacier: the surface warms and the streams of water run to the edge.

Photo: Jan Tove Johansson/Getty Images

Arctic sea ice

The Arctic is experiencing some of the most severe climate change on Earth. Temperatures are rising about 0.5°C every decade, with winter temperatures in western Canada and Alaska 3° to 4°C warmer than 50 years ago. Higher temperatures have reduced the Arctic Ocean's winter ice cover from around 16 million km^2 in the 1960s to only 14 million km^2 by 2006. An area twice the size of Norway, Sweden and Denmark has been lost, half of it since 2001.

Similarly in summer 8 million km^2 of sea ice used to remain frozen, but in 2007 almost 50% of the summer sea ice dramatically melted. For the first time in human memory the Northwest Passage, a short cut from the Pacific to the Atlantic, opened as sea ice disappeared. Additionally at its centre Arctic sea ice is now 40% thinner than in the 1970s, reduced from 4 metres thick to about 2.5 metres, with much of this loss occurring in the last 25 years.

By 2100 up to 50% of the Arctic's average sea ice cover could disappear while late summer sea ice may be lost almost completely even sooner. This would facilitate more oil and gas extraction in the Arctic, increasing fossil fuel supplies and related emissions and so potentially further accelerating ice melt.

Photo: Richard Olsenius/National Geographic Stock/Getty Images

Greenland´s land ice

Greenland is almost four times the size of France and 80% covered by ice 3km thick at the centre. This land ice contains 8% of the world's fresh water. Since the mid-1990s northern Greenland has warmed by about 2.5°C, causing 16% more ice to melt in summer at the edge of the ice sheet than in the late 1970s. In 2005, the worst year on record, 0.7 million km^2 or about 40% of the surface of Greenland's land ice melted. This is now contributing to sea level rise – with a net average 50 to 100 billion tonnes of melt-water lost annually between 1993 and 2003 and even larger losses in 2005.

125,000 years ago polar temperatures were 3° to 5°C warmer than now, much of the ice had melted and contributed to raising sea level, which was 4 to 6 metres higher than today. Towards the end of this century Arctic annual average temperatures could increase by 3.3° to 8.3°C above the 1850–1899 average, with temperatures from December to February rising by 4.8° to 11.9°C. This means temperatures are likely to rise to a similar or even higher temperature than 125,000 years ago. This will cause significant ice loss. The only uncertainty is how fast the ice will melt.

If Greenland's ice were to melt and decrease in volume for a sustained period of a thousand years, the ice sheet would be almost completely lost and would contribute about 7 metres to sea level rise. Greenland's land ice can be expected to decrease continually in volume once the world's average temperature rises in excess of 1.9° to 4.6°C above pre-industrial temperatures or about another 1° to 4°C higher than today.

Aerial view of melt-water lakes and streams on the surface of Greenland Ice Sheet southeast of the Jakobshavn Glacier, Greenland (July 2006).

Photo: James Balog/Aurora Photos/Corbis

Melting permafrost

Permafrost is soil frozen at or below 0°C for two years or more, and for thousands of years in some areas. Amost a quarter of all land in the northern hemisphere is underlaid by permafrost, 1.5km deep in places. In summer the surface layer thaws enabling tree and plant growth. When the plants die in winter, they only partially decompose due to the low temperatures and the permafrost then stores this carbon-rich vegetation.

Since the 1980s the Arctic's upper permafrost layer has warmed up to 3°C, making land unstable in Alaska, Canada and some regions of Russia. Buildings and roads have subsided, trees have toppled over and 125 Siberian lakes have disappeared, believed to have drained as permafrost thawed. Infrastructure such as oil and gas extraction and transportation facilities is also threatened.

The area of permafrost land in the northern hemisphere has already decreased and by 2050 could be reduced by 20% to 35%. As the permafrost thaws, carbon dioxide and methane are released into the atmosphere. 400 billion tonnes of carbon, up to 45 years' worth of current emissions, are stored in the Earth's permafrost. So as temperatures rise, billions of tonnes of greenhouse gases will be emitted, increasing temperatures further in a dangerous self-perpetuating cycle.

Buildings subsiding into each other due to thawing permafrost: Dawson City, Yukon Territory, Canada.

Testimonial: Shismaref, Seeward Peninsula, Arctic

Tony Weyiouanna

For the past 50 years our community has known that the world around us has been changing. We and our grandfathers have noticed the rising water level, thinner ice, warmer winters and summers and shorter springs. Summer and autumn storms are getting stronger and come earlier in the season.

Our 600 people have used the island village of Shishmaref as a winter camp for 4,000 years. Our ancestors followed the seasons, moving from the rivers and streams to the coast and then on to the coastal islands hunting for food – this subsistence lifestyle is still followed. Today we travel by snow machine over the ice and by boat when the ice is no longer safe, hunting for seals and fish. Due to the warmer climate our community is forced to gather our traditional foods earlier in the season. Shishmaref is also facing permafrost melt. Buildings are collapsing and there is flooding. The only solution is to relocate the community off the island to the mainland.

But we have been here for countless generations. We are Shishmaref, we are Inupiaq Natives, subsistence is our way of life. Who and what we are is based on where and how we live. We value our way of life, we value the environment as it sustains us. We are a community tied together by family, common goals, values and traditions. Shishmaref's community has a long proud history, but we are being forced to leave due to climatic changes.

A home destroyed by beach erosion tips over in Shishmaref. Reduced sea ice, exposing the shoreline to waves, sea level rise and thawing permafrost are making the shoreline vulnerable to erosion (September 2006).

Photos: Author picture – Bryan and Cherry Alexander Photography
Main picture – Gabriel Bouys/AFP/Getty Images

The Antarctic

In 2002 the Larsen B Ice Shelf collapsed off the South Pole's Antarctic Peninsula. The vast block of ice approximately 100km long, 95km wide and 200 metres thick was one of five massive ice shelves to break up between 1995 and 2002 as regional temperatures rose rapidly by up to 2.5°C. In 2008 part of the Wilkins Ice Shelf collapsed. A huge iceberg fell away leaving just a narrow ice bridge connecting the ice shelf to two islands – this too collapsed in April 2009. Additionally since the 1950s, 90% of the Antarctic Peninsula's 244 glaciers have been melting.

The Antarctic Peninsula is part of Antarctica, a landmass 25% bigger than Europe and almost completely covered by an ice sheet millions of years old that contains about 70% of the Earth's fresh water. Since the early 1990s, the ice sheet has lost on average a net 150 billion tonnes of melt-water a year, which is contributing to sea level rise.

Next to the peninsula, parts of the West Antarctic Ice Sheet (WAIS) are also melting unexpectedly. Scientists fear the 'giant' has awakened. Just a further 2°C rise in the Earth's average temperature, if sustained, could trigger significant melting of the WAIS. If it were to melt completely it would contribute 5 metres to sea level rise over the next thousand years or more.

Part of the remaining Larsen B ice shelf. On the left lies the front of the ice shelf, on the right an ice mass that broke off to form an iceberg. (March 2008).

Photo: REUTERS/Pedro Skvarca/IAA-DNA/Handout

Over 80% of the heat added to the climate system has been absorbed by the global oceans, which have warmed to depths of at least 3,000 metres.

This warming causes sea water to expand, and contributes to sea level rise.

Aerial photograph of flooded farms in southern Bangladesh.

Photo: Peter Essick/Aurora Photographs

Rising sea levels

Sea level has risen and fallen over thousands of years between warm interglacial periods and ice ages. It rose 120 metres over several thousand years after the last ice age, about 21,000 years ago and is rising again today as warmer ocean waters expand and as glaciers, ice caps and ice sheets melt. During the 20th century global average sea level rose an estimated 17cm at an annual average rate of 1.8mm between 1961 and 2003. From 1993 to 2003 this increased to 3.1mm a year.

As oceans rise coastlines are eroded, coastal wetlands including saltwater marshes and mangroves are inundated, salt waters invade freshwater supplies and land is lost to the sea. In the South Pacific, two uninhabited Kiribati islands were submerged in 1999. In the Maldives, with over 1,000 small islands, 80% of the land is less than 1 metre above sea level.

By 2100 sea level could rise another 18 to 59cm if Greenland and Antarctica's contribution stays at 1993–2003 melt rates. If it grows linearly as the world's temperature increases, sea level could rise up to 79cm. Some scientists even predict at least a 1 metre increase. Just a further 2° to 4°C rise in the Earth's average temperature, if sustained for millennia, would lead to a significant contribution to sea level rise from the Greenland and West Antarctic Ice Sheets, raising sea level up to 12 metres over the next thousand years or more.

Bora-Bora in French Polynesia in the South Pacific.

Testimonial: Tuvalu, island in the South Pacific

Siniva Laupepa

It is scary living in Tuvalu now that changes in the climate are inevitable. First, the frequent storms, that people of Tuvalu are now experiencing give fright, because sometimes houses are damaged, trees are uprooted, sea-gravels are washed and piled on to roads, crop vegetations get spoiled and harmed. These are results of storms and cyclones.

What's more, the rise of water to unusual places is unavoidable. In recent times during high tides the sea water has risen higher and higher and has gotten to places it never reached before, going right up to people's homes which are about 100 metres away from the sea. In these areas vehicles cannot be used. This is something I have never experienced before in Tuvalu. My memory of floods is associated with snapshots I have seen in overseas countries where people hung on to floating items in fear, or climbed on to roof-tops hoping for survival. I never thought that Tuvalu would be flooded but it is happening now and it is scary. Obviously, the change cannot be denied and people have to realise that the situation is getting worse day by day. They should also realise the contributing factors to the problem and most importantly what they need to do to minimise the problem. People who have the means and can afford to have started to leave Tuvalu with the majority ending up in New Zealand because entry into the country is a lot easier than Australia. Yet the population of Tuvalu does not have the money to leave. The hope is that the government will help them.

Tuvalu, Funafuti (February 2002). Waves from the atoll lagoon wash into a family home during exceptionally high tides.

Photo: Mark Lynas/Still Pictures

During the 20th century melting glaciers and ice caps have decreased in volume worldwide, with the rate of ice loss almost doubling since 1990.

This image shows the trimline, the line below the vegetation, of the Columbia glacier in Alaska (1984). By 2006 it had receded by 400m due to ice melt. This loss is equivalent to the height of the Empire State Building.

 Photo: James Balog/Extreme Ice Survey

Approximately 1200 feet

The oceans' heat-transporting currents

As mountain glaciers and ice sheets melt at accelerating rates and as precipitation patterns change, more fresh water flows into the sea, reducing its salinity or salt content. This could slow the Meridianal Overturning Circulation (MOC) – powerful ocean currents transport heat from the equator to the polar regions, where surface waters cooled by cold winds become dense with salt, causing them to sink. The waters then flow back along the bottom of the sea towards the equator. In the North Atlantic, the Gulf Stream's warm current, more properly known as the North Atlantic Drift, provides warmth to the east coast of America, northern Europe and Britain as it travels north to the Arctic, making Europe up to 8°C warmer.

However the oceans' reduced salinity is likely to undermine the sinking process and although the MOC is not expected to stop this century it could slow by 0% to 50% by 2100. As the ocean currents slow the warmth they provide will decrease, although initially the higher temperatures due to global warming are expected to overwhelm the localised cooling associated with a slowing of the MOC. Beyond 2100, should the MOC stop, annual mean temperatures in most of Europe could fall by 2° to 4°C, with a greater cooling in extreme north-western parts.

This image shows the ocean circulation system in the North Atlantic. Warm near-surface water from the Gulf of Mexico is drawn north as cold, dense, salty waters sink in two polar regions and then travel back near the ocean bottom towards the equator.

Gulf Stream

Gulf of
Mexico

Atlantic
Ocean

 Warm surface current

Cold deep current

Convection areas

Acidic oceans

The oceans are a key part of the climate system. They contain about 37,000 billion tonnes of carbon – almost 50 times more than in the atmosphere. An estimated 90 billion tonnes of carbon is exchanged between the ocean and the atmosphere every year as part of the natural carbon cycle.

But now carbon dioxide emissions from human activities are affecting the oceans. The oceans have been absorbing about two billion tonnes of this carbon every year, which is making them more acidic. The increased acidity is expected to affect marine ecosystems – corals and other shell-forming organisms may no longer be able to grow their calcium carbonate shells and some types of plankton at the base of the food chain could be damaged. Coral and plankton numbers could decrease within decades, reducing the fish populations that feed on them.

The higher ocean temperatures are also likely to reduce the amount of carbon dioxide the oceans absorb from the atmosphere. Recently, Antarctica's Southern Ocean, which absorbs an estimated three quarters of all the carbon dioxide taken up by the oceans, looked to be absorbing less. This reduction in ocean uptake of carbon is happening far sooner than climatologists thought possible and means more carbon dioxide will stay in the atmosphere, so accelerating global warming.

Up to one third of the Earth's plant and animal species face extinction if the world's average temperature rises a further 1.5°C.

An infant elephant is given water by local Touareg after being trapped for three days in a well during a desperate attempt by a migrating herd to find water. Climatic variation, and its effect on the reliability and quantity of rainfall, could severely affect the last remaining Sahelian elephants that inhabit the Gourma region of Mali.

Photo: Jake Wall/Save The Elephants www.savetheelephants.org

Ecosystems

The Earth's ecosystems are already under attack from deforestation, pollution and over-exploitation of resources. And now our changing climate looks set to overwhelm them. Ecosystems that have taken millions of years to evolve may be wiped out in a single century.

Already the natural world is changing in response to global warming, with earlier flowering, leaf unfolding, fruit ripening and changes in the distribution of species. Species have also been undermined: 50% of Costa Rica's frogs and toads have become locally extinct, 20% of polar bears in the Arctic's western Hudson Bay have been lost and 80% of Antarctic krill. Currently 35% of the world's bird population, 50% of amphibians and 70% of warm-water corals are endangered.

It is the speed and magnitude of change that is threatening species. Evolution has taken millions of years, but today's warming is demanding species adapt in a few decades. Even 0.5° to 1°C of warming per century is too fast for many species. Small climatic changes can make their environment unsuitable and if they cannot adapt or migrate they will perish. Just a 1°C rise this century will threaten thousands of species with extinction – but temperatures could rise by up to 6°C.

The golden toad, *Bufo periglenes*, lived in a limited region of high altitude cloud-covered tropical forests, above the city of Monteverde, Costa Rica. The last sighting of this small toad was in 1989. Its restricted range, global warming and pollution are believed to have contributed to this species' extinction.

Photo: Michael Patricia Fogden/Minden Pictures/National Geographic Stock

Testimonial: Polar bears on the edge

Andrew E. Derocher, PhD
Former Chairman of the Polar Bear Specialist Group of the World Conservation Union

In the Arctic's western Hudson Bay the number of polar bears has decreased by 20%. By 2050 two thirds of the polar bear population could be lost. They are now threatened with extinction.

Few species are as recognisable as the polar bear – the white bear in a white land. Having evolved from a brown bear ancestry, the polar bear moved into a new habitat devoid of any large predator and made the shifting sea ice of the circumpolar Arctic its home.

Within any given year a single bear can wander over thousands of kilometres. The bears rely on the sea ice for a platform to hunt, mate and migrate. Ponder life at –40°C in the dark days of winter with 24 hours of night – it's not easy for polar bears but they are incredibly successful at what they do.

The Arctic sea ice is rapidly diminishing. This starts a chain reaction for the bears: as the sea ice melts, the bears have a harder time finding seals and with fewer seals, the bears lose body condition. With less stored body fat, pregnant females are less successful at rearing cubs and with fewer cubs, the population declines. The symptoms of climate warming are showing all over the Arctic: bears are getting thinner, reproductive and survival rates are declining, some bears are becoming cannibalistic in their desperation, and others drown due to lack of sea ice. In other areas, the evidence is simpler but equally insidious: the ice is gone and so are the bears. If you take away their habitat, they become extinct. The Arctic would not be the same without the bears – their home is melting from beneath them and time is running out.

Photos: Author picture -- A. E. Derocher
Main picture – Flip Nicklin/Minden Pictures/National Geographic Stock

Dying reefs

Coral reefs are one of the world's major ecosystems, harbouring up to a quarter of all fish species. Already 20% of all reefs have been lost due to coral bleaching and human activities such as destructive fishing practices, tourism and coastal developments. Deforestation also causes greater soil erosion that results in increased sedimentation, which can overwhelm corals. 35% of the world's corals are now under imminent or long-term risk of collapse with climate change the biggest threat – the more acidic oceans and rising sea temperatures could completely wipe them out this century.

Even a 1°C rise in ocean temperatures can cause corals to bleach, or go pale in colour. Once bleached the corals sometimes recover but often die. In 1998, one of the hottest years on record, 16% of the world's corals bleached and in 2002 over 60% of Australia's Great Barrier Reef. By 2050 most of the Great Barrier Reef is expected to bleach annually.

With a further 1.5° to 3°C rise in the Earth's average temperature coral reefs are unlikely to recover. Not only will this damage the US$90 billion-plus fishing industry and the lives of over 40 million people who work in it, but it will also severely affect one billion people who rely on fish as their main source of protein. Most critically one of the Earth's major ecosystems will be lost.

Photo: Gary Bell/Corbis

This is a dead coral reef. An effective illustration of how our world, rich in biodiversity, may be transformed as climate change takes hold.

Forest ecosystems

Another major ecosystem, the world's forests, harbour up to 90% of Earth's species. Every year 13 million hectares are lost through deforestation, with climate change now set to accelerate the toll. Just a 1°C change in annual temperatures can affect a tree's productivity, while pests, forest fires, and droughts may decimate many forests this century.

If the amount of carbon dioxide in the Earth's atmosphere doubles, one third of the world's forests could undergo major changes or entire forests could disappear. The Amazon rainforest, ten times the size of France and home to millions of plant, animal, bird and insect species, is particularly threatened. At least 20% to 40% of the Amazon is expected to be committed to die-back with even a small warming of about 2°C above pre-industrial levels. With a 4° to 5°C increase about 90% of the forest could be gone and all the species that inhabit it – an ecosystem 55 million years old lost in two centuries due to human actions.

Worsening wildfires will also destroy forests – as in 1998 when 9.5 million hectares burned in Indonesia. And as forests die or burn, they no longer absorb carbon dioxide from the atmosphere – instead the carbon they store is released. 1998's wildfires emitted an estimated 0.8 to 2.6 billion tonnes of carbon, or up to 30% of current human emissions. As temperatures rise and as more forests burn, their emissions will further fuel global warming.

Photo: John McColgan, Bureau of Land Management, Alaska Fire Service

Thousands of species threatened

Nature's web has been finely sewn and humankind cannot hope to pull at one thread without the web being broken. A world that existed in perfect balance for hundreds of thousands of years threatens to fall apart degree by degree as global temperatures rise. If carbon dioxide levels in the atmosphere double, as expected by 2050 if we do not act, thousands to tens of thousands of species are predicted to be committed to extinction.

Based on pre-industrial temperatures, with a temperature rise of:

1°C	Arctic ecosystems will be increasingly damaged
1.7°C	corals in South East Asia, the Caribbean and the Great Barrier Reef bleach
2–3°C	the Amazon rainforest dies back
2.3°C	15% to 35% of plant and animal species face extinction
2.8°C	high risk of extinction of polar bears, walruses and seals
3°C	20% to 50% of plant and animal species face extinction
	Above 3°C few ecosystems can adapt.

Once lost, they are gone forever.

In 2008 Arctic skuas, Arctic terns and kittiwakes suffered a near total breeding failure in the Northern Isles of the UK. This is believed to be due to rising sea temperatures reducing the number of sandeels on which the birds feed.

 Photo: Andrew Parkinson/Corbis

A CHANGING WORLD FOR HUMANITY

Humanity is dependent on the natural world and the world's climate is fundamental to our life support. We may have distanced ourselves from Nature, but it remains central to our existence. We all need food, water, shelter, bearable temperatures. Without these basics we will flounder.

Our changing climate now threatens humanity. And although a few impacts may be positive, with fewer deaths from winter cold and increased crop production in some regions with a small warming, most will be negative and severe and will escalate as temperatures rise.

Photo: Daniel Cima/American Red Cross

An estimated 325 million people each year are already severely affected by climate change.

By 2030 the number is expected to more than double to about 660 million a year as temperatures rise and as environments deteriorate.

The capital of Honduras, Tegucigalpa after Hurricane Mitch struck (1998). 11,000 people were reported dead with thousands missing. More than three million people were made homeless.

Flooding

Every year over 125 million people are affected by floods and flash floods – a six-fold increase since the 1970s. 1998, one of the hottest years on record, was a particularly bad year. In China floods affected 240 million people: 4,000 people died, 18 million homes were destroyed or damaged and 20% of arable land was inundated. A year later in Venezuela, devastating floods and landslides caused 30,000 deaths. But if the world's average temperature rises 2° to 3°C between one and five billion people are likely to be at greater risk of flooding.

The Asian monsoon could also be affected as rainfall patterns change. Mainly wetter monsoons are expected, increasing water availability for two billion water-stressed people in South and East Asia. But this will also exacerbate monsoon flooding. Recent events suggest this could be happening already. In 1998 two thirds of Bangladesh was under water for three months and 30 million people were made homeless. Then in 2004 almost 36 million people were flooded. In 2005 India experienced its most intense monsoon rain ever, with 944mm falling in 24 hours in Mumbai and surrounding regions. 1,100 people died. In 2007 South Asia suffered severe floods again, with 30 million people affected.

Mumbai, India (2005). Heavy rains caused flooding up to 3 metres deep. Many slum dwellings were totally destroyed, people were buried as walls collapsed, other were electrocuted or drowned in vehicles.

Testimonial: Floods in Bangladesh

Monimala Sarker from Baikanthapur village

Bangladesh is very vulnerable to annual monsoon floods and storm surges. It is in a delta, densely populated and low-lying with 10% of the land barely above sea level.

I have never experienced anything as severe as the floods of 1988 and 1998. The floodwaters inundated my homestead in 1998. My husband made a tong, a platform, out of bamboo and hyacinth plant. We then moved two roofs from our home to the floating tong, which was only a few square yards. On this small area we lived for nearly a month, surrounded by floodwaters.

Children drowning was a major risk. We tied our two boys to us or the tong all day and night. My six daughters, aged 15 to 24, spent the days in the branches of trees, as the tong was too weak for all of us. At night they slept on the tong and next morning would return to the trees where they ate, drank and bathed. Increased numbers of snakes, frogs, mosquitoes and other insects made our lives horrible.

Cooking was a major problem. I used a small portable oven on the tong for cooking but there was little rice and food so some days we starved. Water was another acute problem. All wells and pumps were under water so we used floodwaters for drinking and cooking. We suffered from bad headaches, coughs, allergies, itchiness and skin diseases. Many villagers suffered diarrhoea, dysentery and skin ulcers, particularly on their legs and feet. Our suffering became worse after the flood since the main resources, paddy fields and household items, were damaged. We became asset-less and money-less and unable to maintain our livelihoods without financial assistance or aid from outside.

Photo: REUTERS/Rafiquar Rahman

With a 3° to 4°C increase in the Earth's average temperature 330 million people could be displaced worldwide due to flooding and sea level rise – over 70 million in Bangladesh, six million in Lower Egypt and 22 million in Vietnam.

Photo: REUTERS/Rupak De Chowdhuri

US$7 trillion worth of insured US property is currently at risk from North Atlantic hurricanes, including 60% of property in Connecticut and New York and 80% of property in Florida.

Over 35 million Americans living in coastal communities are now at risk.

Neighbourhoods flooded with oil and water after Hurricane Katrina hit New Orleans (August 2005).

Coastal communities threatened

Human settlements in both developed and developing regions are likely to be increasingly endangered by changes in tropical storms and storm surges. A storm surge is a wall of water pushed towards the shore as a result of strong winds and low pressure. Hurricane Katrina's storm surge produced waves up to 8.5 metres high along the Missisipi coast and 5-metre waves that breached the levees of New Orleans. Many cities worldwide are at risk, including New York, Tokyo, Hong Kong, Shanghai, Mumbai, Buenos Aires and London.

Sea level rise will also inundate tens of millions of people. More and more people are moving to live in coastal flood zones, mainly in urban environments. This together with sea level rise could increase the number at risk of coastal flooding by ten-fold or more, to over 100 million people a year later this century.

Major cities are in danger including Bangkok, New Orleans and Shanghai, along with small islands and major delta regions in Bangladesh, China and Africa. In China over 30 million people live below the 50cm water line and as much as 30% of Africa's coastal infrastructure could be submerged this century.

This computer generated image from the film *'Flood'* shows the impact of a storm surge wave overwhelming the London Thames Barrier and flooding London.

Photo: Computer generated image 'Courtesy of Power'

By 2030, almost half of the world's population will be living in areas of high water-stress due to climate change and unsustainable water use.

Most of these will people will come from developing countries.

In China 400 major cities are already critically short of water, with people restricted to one quarter of the world average. By 2050 China's population could increase by 250 million, further reducing the amount of water available per person.

Water resources

Water is the source of life; but each person alive today has only 40% of the water that was available per person in the 1950s, largely due to population growth. Currently up to two billion people on Earth live in severely water-stressed basins and every year an average 110 million people are affected by drought. As economies develop and as the global population expands to over nine billion by 2050, water demand could increase a further 30% to 85%. This will mean even less water per capita.

About 90% of the population growth will occur in developing countries, where in many regions clean water is already severely lacking. Escalating water withdrawals will cause 62% to 76% of Earth's land areas to become increasingly water-stressed. At the same time water supplies will diminish as climate change increases droughts and further reduces water availability in dry regions such as the Mediterranean basin, the western US and southern Africa. Rising oceans will cause salt water intrusion into freshwater supplies and melted glaciers will affect water supplies for up to one billion people.

With just a 2°C rise in temperature up to four billion people could suffer increased water shortages. In southern California 40% of water supplies could be threatened by 2020 and in India, the amount of water per person could decrease by 40% by 2050.

Photo: International Federation of Red Cross and Red Crescent Societies

Drought in Australia

Australia is already suffering from a lack of water. Temperatures have increased 0.75°C in the last 15 years and rainfall has decreased, leading to water scarcity and drought. Since 2001 the Murray Darling river basin in southeast Australia has had the worst multi-year drought since 1900. This region is home to 1.8 million Australians, grows 70% of the country's irrigated crops and provides 85% of the water used nationally for irrigation. In 2002–2003 the drought caused a 1% reduction in the country's GDP and the loss of a 100,000 jobs. In 2003, grain yields decreased by 50%, millions of sheep and cattle died and over 80% of dairy farmers were affected.

In 2006–2007 there was still only minimal water available for agriculture and Australia's rice crop was reduced by 90%. Farmers were ruined – unable to make a living in what was now a dustbowl, many sold up and abandoned land that their families had farmed for generations. Others committed suicide. Long term the outlook appears increasingly bleak. Decreased streamflow is likely over most of southern and sub-tropical Australia, which will increase drought and wildfires, further reducing crop production.

Melting glaciers

One billion people in India, China and the Andes depend on melt-water from glaciers or snow-fed rivers for their dry season water supply. This century they could lose this source of water as glaciers melt and then disappear completely.

Initially water flows will increase, as rising temperatures melt glaciers and snow more rapidly and earlier in spring. This could cause glacial lakes to burst or overflow, as in Bhutan in 1994 when 21 people died. In the Himalayas dozens of lakes are close to bursting, threatening thousands of lives, property and infrastructure. Then once the glaciers and snow have melted completely this water source will disappear, threatening hundreds of millions of people who rely on it for domestic and agricultural purposes.

In the Andes 50 million people rely on glacial melt-water during the dry season, but the land covered by glaciers has decreased by 25% in the last 30 years. In the Himalayas melting glaciers could affect water resources in the next 20 to 30 years. The melt-water provides 70% of the Ganges summer flow, which supplies 500 million people with water. In China most glaciers are retreating, with two thirds of them expected to disappear by 2060 and all by 2100, cutting off the water supply to over 250 million people.

People crowd on the banks of the river Ganges in Patna, India.

Every year nearly six million children die before their fifth birthday due to hunger or lack of nutrition.

The majority are children born in developing countries.

75% of all these child deaths are in sub-Saharan Africa and Asia, where conditions are expected to deteriorate further as temperatures rise.

Food production

Today over 920 million people are undernourished globally, 80 million more than in the early 1990s, with 90% from Africa and Asia and one quarter children. Every day 25,000 adults and children die from hunger or hunger-related diseases. Climate change is already considered to be the root cause of hunger and malnutrition for about 45 million of these undernourished people. By 2030 this figure could rise to 75 million.

Initially however, climate change could improve crop production for countries beyond the tropics as local temperatures rise 1° to 3°C. Northern Europe, North America, and east and south-east China could all see yields increase, although floods, droughts or heat stress could undermine potential gains. But in seasonally dry and tropical regions, primarily home to developing nations, even a 1° to 2°C local warming will cause crops to decline. In most Asian countries yields have already decreased, due partly to rising temperatures and a reduction in the number of rainy days. In the tropics crops are threatened above 35°C, and over 40°C few can grow.

If temperatures increase by over 3°C, all crops in all regions are likely to be negatively affected. With a 3° to 5°C rise major cereal crops, such as wheat, maize and rice, could see losses of up to 60%. The highest losses will be in tropical regions where it is not possible to adapt by using different seed varieties, changing planting times and shifting from rain-fed to irrigated agriculture.

A field of wheat submerged by floodwaters in England (July 2007). Approximately 42,000 hectares of agricultural land was flooded across England and 2600 to 5000 farms were affected.

Failing crops

By 2050, developing countries could see food production decline dramatically. In central and southern Asia crop yields could decrease by up to 30% and in some regions of Africa up to 50% of rain-fed crops could fail by 2020. With a 4°C rise parts of Australia and Russia could be completely out of production and at 5° to 6°C agricultural collapse may occur worldwide.

Livelihoods would be devastated. 45% of the global workforce works in agriculture, including 75% of the poorest people globally – the one billion living on less than US$1 a day. In India and China 60% of working people work in agriculture and in Africa 70%. In stark contrast only 2% of workers in the UK and US are employed in this sector.

By 2050 demand for food could increase by 70% to 80% as the population rises to over nine billion. The number of people who will go hungry as the world warms is expected to depend on economic development, the carbon fertilisation effect and weather-related disasters. Economic development will raise living standards and could reduce the number undernourished, while the extra carbon dioxide in the atmosphere could 'fertilise' crops, so boosting their growth before yields decline with greater warming. However increased floods, droughts and higher temperatures could decimate crops and threaten livestock.

A farmer examines part of his parched field near Chandigarh, India (July 2002).

Photo: Stringer/AFP/Getty Images

'Testimonial: Drought in Malawi

Chrissie Nyirenda

In 2002 Malawi suffered the worst maize shortage in living memory. Hundreds of people died and three million people faced starvation. In 2005 heavy rains followed by drought caused severe crop losses again – this time five million were short of food. Since then above average winter rainfall has returned to Malawi but long term the outlook is poor. Rainfall and water availability are likely to decline as temperatures increase 2° to 3°C by 2050.

Malawi used to get a lot of rain, but since 1998 it has been struck by droughts. November to December used to have temperatures of 30° to 40°C but now these temperatures come any time. Rains come a month or two later, crops have to be planted later, leading to a delayed harvest which can cause famine.

In 2002 drought left us with a poor harvest and the price of a bag of maize, our staple food, went up by over 100%. An ordinary Malawian could not afford that and had to survive on hand-outs. People died of hunger and three million faced starvation. Some livestock, particularly cattle, died as the grass and small rivers dried up. In 2005 another drought forced a bag of maize to increase from US$7 to US$19. An average Malawian survives on less than a US$1 per day and families could not cope. Over 4.5 million people were short of food and the Malawi government spent over US$100 million buying maize to save lives. Several donors had to assist as well. Some people died but most were saved through the Government's intervention. During the drought there was little available water so people walked over 5km daily to fetch it.

The Malawi climate has definitely changed and an ordinary Malawian does not understand why. "

Photo: International Federation of Red Cross and Red Crescent Societies

At risk of hunger

Studies have sought to estimate how many people will be at risk of hunger this century as the world warms. With economic development and with extra carbon dioxide in the atmosphere 'fertilising' crops in some regions, the number could fall from 920 million today to 100–380 million by 2080. But with high greenhouse gas emissions 400 to 600 million more people go hungry. Sub-Saharan Africa will be particularly hard hit, accounting for 75% of those at risk of hunger.

However these statistics could well be underestimates, as most studies to date largely exclude the spread of pests and diseases and the effect of increasing floods, droughts and windstorms on agriculture – yet their effect could be substantial. Europe's 2003 heatwave reduced maize yields in France by 30% and in Italy's Po Valley by 36%, while Brazil's 2004–2006 drought reduced soybean and maize production by up to 65% and 56% respectively. In Ethiopia the drought of 1998–1999 killed 62% of all cattle.

Additionally, more than 80% of global agricultural land and almost 100% of pasture land is 'rain-fed', so that changing rainfall patterns could have severe consequences. Already a quarter of the Earth's land surface is too dry for rain-fed agriculture and if areas in extreme drought increase from 1% to 30% of all land areas by 2100, crop yields could be severely affected. The impact of weather-related disasters, the replacement of food crops with biofuel crops and higher food prices due to reduced supplies may severely exacerbate food shortages. These could all combine to increase famine, malnutrition and death.

Since the 1970s windstorm disasters have increased three-fold, drought disasters have increased five-fold, flood disasters have increased six-fold and the number of people affected by natural disasters has tripled to an average 280 million people a year.

Stevenson Ranch (2003) southern California. 11,000 firefighters battled one of southern California's worst ever wildfires. 22 people died and 3,600 homes were destroyed.

Disasters

The number of natural disasters the world faces today is unprecedented.

Natural disasters have quadrupled since the 1970s and although there are fewer deaths, the number of people affected has tripled to an average 280 million people a year. Of those affected 95% come from Asia and Africa and when disasters strike these poorer nations, people can lose all they have.

The cost of the damage from weather-related disasters has also grown dramatically. Economic losses and insured losses have risen seven- and 29-fold respectively to US$48 billion and US$22 billion a year * since the 1960s. 2004 and 2005, the worst hurricane seasons on record, produced economic losses of US$114 billion and US$230 billion respectively. The latter amounted to 0.5% of global Gross Domestic Product (GDP).

The growth in financial losses and in the number of people affected by disasters is partly due to population growth, increased wealth and people living in more vulnerable areas and partly due to climate change causing more extreme weather events. By 2050, annual weather-related disaster losses could cost 0.5% to 1% of world GDP – up to US$1 trillion a year at a conservative estimate. And by 2100 losses could reach several per cent of GDP if greenhouse gas emissions continue to increase this century.

* at 2008 values.

An aerial view of flooded New Orleans, Louisiana following Hurricane Katrina and the breaching of the city's sea defences by storm surge waves up to 8.5 metres high (August 2005).

Photo: REUTERS/Vincent La Forêt

Fatalities

Climate change already causes an estimated 300,000 deaths a year, largely due to weather-related disasters and environmental deterioration. As the Earth continues to warm, the number of climate change related deaths is expected to rise to almost 500,000 deaths a year by 2030.

Additionally as natural disasters worsen and become more frequent, Gross Domestic Product (GDP) will decrease. This will initially be seen in poorer nations, then later also in developed countries as climate change takes hold. In less developed countries this is likely to lead to greater poverty and more childhood deaths – a 5% loss in a country's GDP could increase infant mortality by 2%. By 2100 this could mean 40,000 to 250,000 more children dying each year in sub-Saharan Africa and South Asia due to increased poverty caused by climate change alone.

A child suffering from dengue fever, at Phnom Penh's Kantha Bopha VI hospital (July 2007). By the 2080s the number of people at risk of dengue fever is expected to increase from 1.5 billion today to 5–6 billion people, with 1.5 – 2.5 billion attributable to climate change.

Water wars

The impacts of climate change, combined with population growth and political or social unrest, could lead to national and cross-border conflicts, particularly concerning water scarcity. West Africa, Central Asia and the Nile region are all potential risk zones.

Ten countries use and need the waters of the Nile: Sudan, Kenya, Uganda, Burundi, Tanzania, Rwanda, the Democratic Republic of Congo, Eritrea, Egypt and Ethiopia. By 2050 the population of these countries is expected to more than double, from 410 million people in 2009 to 860 million. At the same time, lower Nile river flows could decrease by more than 60% by 2050 and by as much as 75% by 2100, due to potential reductions in rainfall, increased evaporation caused by higher temperatures and escalating water withdrawals due to rapid population growth. This will cause a water crisis.

"
The next war in the Middle East will be fought over water, not politics.

Former UN Secretary General, Boutros Boutros-Ghali (1985)
"

Photo: REUTERS/Finbarr O'Reilly

By 2050 mass international migrations are expected, with tens to hundreds of millions of climate change refugees – people fleeing their homelands searching for food, water and shelter or escaping rising oceans.

Where will all these people go?

Photo: REUTERS/Ho New

Developing countries

"The effects of climate change are expected to be greatest in developing countries in terms of loss of life, and relative effects on investment and the economy ... Those with the least resources have the least capacity to adapt and are the most vulnerable.

Intergovernmental Panel on Climate Change (2001)"

Climate change will increasingly overwhelm efforts to make poverty history, by making the lives of those in developing countries largely much worse. It will increase heat-stress, disease and disasters, affect food and water resources, and undermine livelihoods, incomes and future growth prospects of some of the least developed countries and the 2.6 billion people who live on less than US$2 a day. Those that are the most vulnerable will suffer the most, even though they are the least responsible.

Developed nations may, on the other hand, initially experience some benefits, with improved crop yields, fewer deaths from winter cold and increased tourism as regions warm. Such benefits will however be followed by predominantly negative effects long term, but these will pale alongside the impacts on poorer nations. Unless the solutions to climate change are adopted worldwide, with developed countries providing significant financial support to developing nations, some of the most vulnerable people in our world today will have little chance of avoiding an ever downward cycle of poverty and suffering.

Photo: REUTERS/Rafiquer Rahman

The dimming of the planet

Ironically, sulphate aerosols produced by burning fossil fuels have locally shielded regions of the Earth from up to 50% of the sun's rays. If it were not for these aerosols, the world would be even hotter than it is today. The downside is that aerosols cause air pollution and reduce air quality. Sulphate aerosols may also have caused a reduced warming of the northern tropical oceans, which is thought to have led to decreased rainfall in the Sahel region of Africa, leading to drought and severe famine in the late 1960s to the early 1980s.

Between 1980 and 2000 however, global emissions of sulphate aerosols decreased by about 25% as countries, particularly developed nations, sought to reduce these pollutants. The diminishing shielding effect of the aerosols may have contributed to rising temperatures, coupled with increased greenhouse gas emissions.

And as we clean up pollution the shielding effect will be further reduced, making our planet warmer still. The UN's Intergovernmental Panel on Climate Change affirms that action aimed at keeping the Earth's average temperature below a certain threshold must not only take into account carbon dioxide emissions but also measures undertaken to reduce air pollution.

Air pollution above Mexico City.

Photo: Julio Etchart/Still Pictures

Critical feedback

By 2030, trees, vegetation and soils are likely to absorb less carbon dioxide from the Earth's atmosphere. As temperatures continue to increase, forests and soils are expected to release escalating amounts of the carbon they store. Up to 240 billion tonnes of carbon could be released by 2100, or 25 years' worth of current emissions. Already tundra, permafrost and some soils are releasing greenhouse gases, with emission rates accelerating in the last 20 years. Warming oceans are also likely to absorb less carbon dioxide, leaving more in the atmosphere.

All this is critical, as Nature's carbon release will increase global warming, causing these sources to release yet more carbon in a self-perpetuating cycle known as a positive feedback. Up to 2,000 billion tonnes of methane could also enter Earth's atmosphere due to the melting of methane hydrates – an ice-like mix of methane stored deep in the ocean's seabed or closer to the surface in the Arctic.

To put it simply, if we do not rapidly reduce our own greenhouse gas emissions – and soon – we will reach a 'tipping point' which triggers the release of vast amounts of greenhouse gases from Nature's own sources.

If this occurs there will be nothing we can do to stop it.

Photo: Fry Design Ltd/Getty Images

The changes that are occurring to the Earth today have been caused by a relatively small increase in the global average temperature of about 0.8°C.

By 2100 the Earth's average temperature may rise by 1.6° to 6.9°C above 1850–1899's level.

And as temperatures rise, impacts will worsen.

Our children and the coming generation will grow up in a much changing world.
They will witness the impacts of a disrupted climate and see suffering on an unprecedented scale.

Once the predicted changes happen it will not be possible to turn the clock back.

What can be done to prevent this?

SOLUTIONS

If humanity continues on its current path, our world will be irretrievably altered this century.
Yet the technologies exist to prevent this.

But we must act now if we are to avoid catastrophe.

Schott Parabolic Trough Mirrors, harnessing the power of the sun.

 Photo: Schott Solar AG

International agreement

In 1990 the United Nations Intergovernmental Panel on Climate Change, the leading body of world climate scientists, affirmed that global emissions would need to be reduced immediately by 60% to 80% below 1990 levels if the amount of carbon dioxide in Earth's atmosphere were to be 'stabilised' at 1990 levels. In 1992 the United Nations Framework Convention on Climate Change (UNFCCC) was signed with the objective to achieve:

...stabilization of greenhouse gas concentrations in the atmosphere at a level that would prevent dangerous anthropogenic interference with the climate system. Such a level should be achieved within a time-frame sufficient to allow ecosystems to adapt naturally to climate change, to ensure that food production is not threatened and to enable economic development to proceed in a sustainable manner.

192 countries have now ratified the Convention. In 2005 the Kyoto Treaty came into force, committing industrialised countries to reducing their collective greenhouse gas emissions by at least 5% below the 1990 level by 2008–2012. National targets were set to achieve this; however many countries are currently off track to meet their targets. A major shift is urgently needed, based on a new global treaty to be formulated in Copenhagen in December 2009 at a meeting of the parties of the UNFCCC.

Plenary meeting during the UNFCCC Bonn Climate Change Talks (June 2009).

Photo: Jan Golinski/UNFCCC

A global target

Under the United Nations Framework Convention on Climate Change, 192 countries have committed to stabilise greenhouse gases in our atmosphere at a level that will prevent dangerous changes to our climate and so our planet. During the UK G8 Presidency in 2005, a conference at the UK Meteorological Office's Hadley Centre considered what is a 'dangerous level of climate change' and how to avoid it.

The world's leading scientists came together to express their views. They stated that the higher the Earth's average temperature rises, the more dangerous the consequences will be. The breakdown on the right shows the relative dangers for specific increases in temperature over and above their pre-industrial level.

Where would you place the target?

Photo: NASA Goddard Space Flight Center Image by Reto Stöckli (land surface, shallow water, clouds). Enhancements by Robert Simmon (ocean color, compositing, 3D globes, animation). Data and technical support: MODIS Land Group; MODIS Science Data Support Team; MODIS Atmosphere Group; MODIS Ocean Group Additional data: USGS EROS Data Center (topography); USGS Terrestrial Remote Sensing Flagstaff Field Center (Antarctica); Defense Meteorological Satellite Program (city lights).

Just some of the impacts as the global average temperature increases above its pre-industrial level:

At 3°C and above
- Greenland and West Antarctic Ice Sheets melt significantly, raising sea level up to 12 metres
- up to 50% of species face extinction
- wheat, maize and rice yields fall by 20% to 60% globally
- up to 400–600 million more people at risk of hunger
- one to six billion people short of water
- 100 million people suffer coastal flooding

At 2°C
- the Amazon rainforest dies at 2°–3°C, releasing its carbon
- significant melting of Greenland ice sheet
- up to one third of species face extinction
- food prices increase up to 20%, poor farmers' incomes drop globally

At 1.5°C
- some coral reefs preserved
- one to three billion people suffer increased water shortages
- 200 million to five billion people receive more water, increasing the risk of flooding
- accelerated ice melt but Greenland and West Antarctic Ice Sheets unlikely to melt irreversibly

At 1°C
- increased droughts, heatwaves, floods, category 4 and 5 hurricanes and wildfires
- ice sheets melting, oceans acidifying, sea level rising
- up to 60 million extra people at risk of hunger, 300,000–500,000 climate-related deaths a year
- glaciers in the Andes disappear, 50 million people lose this water source

What temperature target?

Some scientists at the Hadley Centre conference in 2005 considered the world's average temperature must rise no higher than 2°C above its pre-industrial level. Others considered this was still too high – a third of species would face extinction, Greenland's ice sheet could melt significantly and the Amazon rainforest could die back, releasing its stored carbon. As forests and soils release their carbon, temperatures will rise still higher.

The EU has now reduced its temperature target from 2°C to less than 2°C while the UN's Intergovernmental Panel on Climate Change identifies 1.75° to 2.75°C above pre-industrial levels as the threshold at which the risk of major impacts increases. Others have called for a limit of 1.5°C. This may still be possible if global emissions stop increasing or 'peak' by 2015 or before and if the Earth´s sensitivity to increasing levels of carbon dioxide is less than estimated. Once emissions peak they must then decrease to zero.

" …only in the case of essentially complete elimination of emissions can the atmospheric concentration of CO_2 ultimately be stabilized at a constant level.

Intergovernmental Panel on Climate Change (2007) "

But there is a critical proviso – emissions can only be eliminated completely if major carbon release from forests and soils has not already been triggered.

NASA satellite image of California wild fires showing smoke rising from wildfires spreading along the southern California coastline (October 2003).

Photo: NASA/epa/Corbis

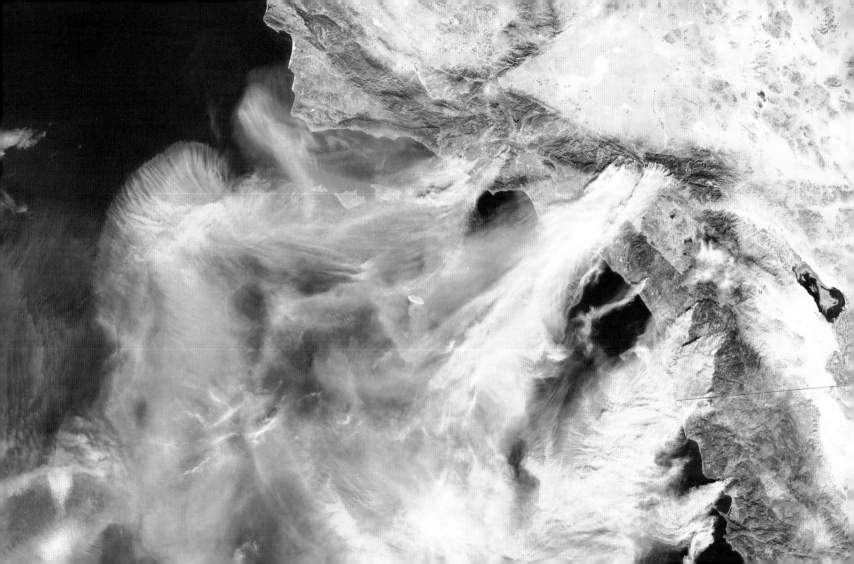

Global carbon dioxide emissions must stop increasing or 'peak' by 2015 or before and then decrease to near zero, if major damage to the Earth is to be avoided.

Wildfires in Malibu, California burning out of control down a hillside towards homes and businesses along the Pacific Coast Highway.

The target

To prevent catastrophic and irreversible damage to our planet, urgent action is needed towards a global target: Aiming to keep the Earth's average temperature rise to 1.5°C above pre-industrial levels. To achieve this:

- Global carbon dioxide emissions must 'peak' by 2015 or before and then reduce by 80% to 90% below 1990 levels by 2050, decreasing thereafter to near zero.

- Developed nations should aim for a near zero–carbon economy by 2050 to allow for economic growth in developing nations

Such changes require a global energy revolution – the rapid transfer from a high carbon to a low carbon energy system – now. And the next few years are the most critical. For if effective action is not taken, the changes to our climate and our world are likely to be irreversible over human timescales. A hotter world, ice sheet collapse and sea level rise are changes that will endure for many thousands of years. Other changes will be completely irreversible – once species become extinct the damage can never be reversed.

This diagram shows how the world is expected to warm under different scenarios relative to the 1980–1999 average. In B1 and A1B the global population peaks by 2050 and then declines – B1 uses clean and efficient energy technologies and A1 uses a mix of energy sources. In A2 the global population increases continuously and technological change is much slower.

Photo: IPCC (refer to references)

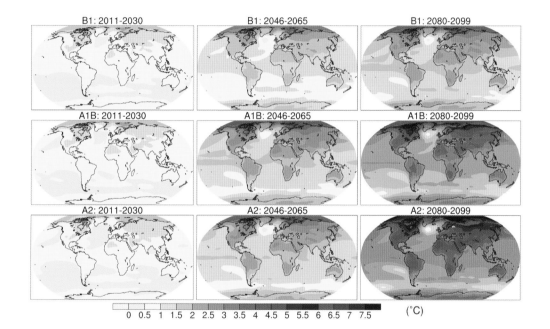

Urgent action

The challenge is considerable. Global carbon dioxide emissions must decrease by 80% to 90% below 1990 levels by 2050, yet are forecast to rise 130% on 2005 levels as the global economy grows four-fold. North America and Europe are the most responsible for climate change to date, having produced about 70% of energy-related carbon dioxide emissions. Developing countries have produced less than 25%. But now up to two thirds of the increase in energy-related carbon dioxide emissions to 2030 is forecast to come from developing nations as their economies and populations expand.

China and India will account for two thirds of this increase. For the last 20 years China's economy has grown about 10% per annum, doubling its growth in the last decade alone. In recent years India has also averaged about 8% growth. As economies grow they use more energy. China's coal use is forecast to double by 2020 – and coal produces the most emissions of all the fossil fuels we burn.

The world population is also exploding, with another two billion forecast by 2050 – more people, more energy and more greenhouse gases. Still worse, the UN Intergovernmental Panel on Climate Change has confirmed that with current climate change mitigation policies, global greenhouse gas emissions will continue to grow over the next few decades.

Shanghai, China (January 2007): Energy consumption is growing rapidly as China´s economy grows at a ferocious pace.
Photo: Fritz Hoffmann/Corbis

Emergency action

A global plan of emergency action is needed to protect the Earth – with the aim of limiting the world's average temperature rise to 1.5°C above pre-industrial temperatures. Global carbon dioxide emissions must 'peak' by 2015 or before and be reduced 80% to 90% below 1990 levels by 2050. Such a plan should include the following measures:

- Put a high price on carbon

- Reduce projected global energy consumption by up to 50% by 2050

- Procure 50% to 60% of global energy from renewable energy sources by 2050

- Capture and store emissions from fossil fuel energy sources

- Prohibit new coal-fired power stations unless emissions are captured and stored

- Prevent deforestation

- 'Draw down' carbon from the atmosphere through improved agriculture and forestry practices

- Enforce the commitment made by developed nations under the United Nations Framework Convention on Climate Change to transfer adequate funds, technology and skills to developing nations.

Photo: NASA/Roger Ressmeyer/Corbis

Strong political leadership and policies

"Preventing catastrophic and irreversible damage to the global climate ultimately requires a major decarbonisation of the world energy sources… Securing energy supplies and speeding up the transition to a low carbon energy system both call for radical action by governments...

World Energy Outlook
© Organisation for Economic Cooperation and Development OECD/International Energy Agency 2008"

Bold political leadership, international collaboration, a dramatic shift in policies and adequate funding are needed to meet the target in time. A long-term market-based policy framework extending to 2050 is also required to secure investor confidence and create market opportunities for low-carbon goods. And with a high long-term price on carbon, by way of a 'climate tax' or cap and trade programmes, significant emissions reductions could be achieved. Carbon could be widely traded with big business acquiring emission permits that can be bought or sold, so generating billions of dollars for low-carbon goods.

World energy subsidies of US$310 billion a year still largely back the fossil fuel 'problem', but must be transferred to support the 'solutions' – removing such subsidies could cut global carbon dioxide emissions by over 6%. And governments must take the lead. Public authorities generally produce 10% to 20% of national emissions. If governments become carbon neutral the cost of low-carbon goods would fall through economies of scale, encouraging investment.

US President Barack Obama and German Chancellor Angela Merkel give a joint press conference in Washington DC, USA (June 2009). The leaders met to discuss various matters including climate change prior to the G8 Summit in July.

Photo: Matthew Cavanaugh/epa/Corbis

If all electrical appliances in OECD countries since 2005 met the best efficiency standards, the emissions saved by 2030 would equate to shutting 400 gas-fired power stations.

A Samsung Electronics LED television. LED televisions can use over 40% less energy than similar sized Liquid-Crystal Display (LCD) televisions.

Photo: REUTERS/Jo Yong-Hak

Energy efficiency

By 2030 the world is expected to use almost 50% more energy than today. But if energy is used more efficiently, less is needed and emissions are reduced. Major emissions reductions are achievable at low cost and quickly through energy-efficient products and buildings, and energy efficiency in transport and industry.

Two thirds of electricity used in industry drives motors; using more efficient motors and other technology can save 40%. Efficient domestic appliances use up to 30% less electricity and even as much as 80% less with advanced models. Take it further – a third of global electricity could be saved overnight if we all changed to low-energy light bulbs. And all of this saves us money!

Buildings can also be made 70% more efficient or even carbon neutral by using modern building technology. BedZED, shown opposite, is the UK's first carbon-neutral village. By 2016 the UK target is for all new UK houses to be similarly carbon neutral – using renewable energy sources to power their heating, cooling and lighting and designed smartly with optimum efficiency. These standards should be adopted internationally. Strict global energy efficiency regulations and standards are needed and energy inefficient products should be phased out. Mandatory labelling of appliances is also essential to increase consumer awareness of energy consumption and efficiency gains.

Beddington Zero Energy Development (BedZED) - smart passive design uses the fabric of the building to warm it and to keep it cool.

Transport emissions

Transport is responsible for almost a quarter of world energy-related carbon dioxide emissions and is expected to contribute one third of the increase in global emissions to 2030. In the next two decades the number of light commercial vehicles worldwide could increase from 650 million in 2005 to 1.4 billion.

Transport emissions must be reduced urgently – by using fuel-efficient engines, low rolling-resistance tyres, lighter, low air-resistant materials and by running cars on diesel or sustainably produced biofuels. Inefficient cars should be heavily road taxed and stringent fuel economy and carbon dioxide emissions standards should be imposed globally following the European Union (EU)'s example. The EU has set an average emissions standard of 130g of carbon dioxide per kilometre for new cars from 2012, with 100% compliance by 2015. This will be reduced further to 95g per kilometre by 2020.

Hybrid cars running on electricity and petrol, and electric cars where long term the electricity is generated from renewable sources, must replace petrol vehicles. While fuel cell vehicles running on carbon-neutral hydrogen could reduce emissions by nearly 100%. Car use can also be reduced by providing cheap, readily available public transport, by promoting cycling and walking and by re-planning how we work and shop. In European, North America and South America cities, bus rapid transit systems – frequent, high speed, efficient buses – have reduced emissions and could be deployed worldwide.

Photo: Spencer Platt/Getty Images

Aviation

Aviation is the fastest growing transport mode. Passenger numbers rose 30% between 2004 and 2007 with emissions increasing at 6% per annum. In 15 years global passenger numbers could double and by 2030 the number of commercial aircraft is expected to rise from 18,000 in 2006 to 44,000.

Over the next 20 years three quarters of the current fleet should be retired; the new aircraft could be up to 50% more efficient due to improved design and better air traffic management. However, an international aviation sector agreement is needed to set standards to ensure this is so. But gains in aircraft efficiency will only partially offset overall growth. Also, aviation is expected to remain wholly dependent on oil-based fuels to 2030.

Voluntary decisions to fly less, to buy fewer products transported long distances and to use video conferencing instead of face-to face business meetings abroad can help. So too will the European Union (EU)'s decision to include aviation in the EU Emissions Trading Scheme. But consumer demand will remain high while flying is cheap because international aviation fuel carries no tax.

To limit flying a high 'climate levy' on aviation fuel for international flights is needed or an 'en route' emissions charge to make air travel more expensive. Revenues raised should be applied to help developing countries adapt to climate change.

Most importantly, aviation must be included in the new climate treaty, to prevent air transport emissions escalating.

Photo: Peter Cross/Monsoon/Photolibrary/Corbis

Renewable energy sources

Reducing energy consumption through the more efficient use of energy means less energy is needed. And this reduction in energy demand will help us to meet the target of securing 50% to 60% of global energy from renewable energy sources by 2050. The technologies exist and the world must transfer from archaic, destructive and dirty energy products to clean sources, harnessing Nature's energy to power our world without destroying it.

Global renewable energy targets should be set every decade towards the goal of procuring 50% to 60% of global energy from renewable sources by 2050. And this means energy targets, not just electricity. Electricity accounts for only one third of global energy and about 40% of energy-related carbon dioxide emissions. So a target to procure 10% of a nation's electricity from renewable sources only equates to about 3% to 4% of total energy. Therefore an overall energy target is needed.

Some countries have set renewable energy targets along the lines of what is needed – China is aiming to obtain 20% of total energy from renewable sources by 2020, equivalent to the European Union (EU)'s firm target. Germany's goal is for at least 50% of energy to come from renewables by 2050, while the EU is pushing for an increase to 60% by 2050. The rest of the world must follow suit, with commitments under a new global agreement.

THE TECHNOLOGIES EXIST TO POWER AND PROTECT OUR WORLD

Over 50% of the world's total energy could come from renewable energy sources by 2050, with renewables capable of providing all of the world's energy many times over.

SOLAR ENERGY Every year the sun's rays provide 10,000 times as much energy as humans need, but solar power currently supplies just 0.04%* of global energy. Solar photovoltaic power systems cover 400,000 rooftops in Japan, Germany and the US and 2.4 million in developing countries. Other solar technologies include solar thermal systems that heat water and space in buildings and industry and concentrated solar power (CSP) that could provide all the world's electricity.

WIND POWER Wind power provides 0.064%* of global energy. Over 125,000 turbines now operate in more than 70 countries. By 2050 wind power could provide up to 30% of the world's electricity and long term much more.

HYDROPOWER Hydropower – energy from flowing or falling water – provides 2% of global energy. By 2025 it could generate up to 10%, rising to 20% long term. It supplies about 60% of Canada's electricity, almost 90% of Brazil's and almost all of Norway's and Iceland's. However, large schemes can have negative environmental impacts.

GEOTHERMAL Geothermal energy – energy from the Earth´s inner heat – provides 0.414%* of global energy but could generate several per cent long term. 40 countries use it to provide electricity and 76 for heating, with over two million geothermal heat pumps in operation globally. 85% of Iceland's heating is supplied by geothermal energy.

BIOMASS Biomass provides 10% of global energy, using energy crops and trees as well as other natural sources to produce electricity heat, fuels, methane or hydrogen. It provides power to over 2.5 billion people in developing countries using traditional wood-burning stoves.

BIOFUELS Biofuels supply about 1.5% of global road transport fuels: ethanol from sugar cane and biodiesel from plant oils. By 2020 10% of European Union transport fuels are targeted to come from biofuels. However a system of certification and monitoring is needed to ensure such fuels are sustainably produced so as to preserve food production and forest cover.

OTHER TECHNOLOGIES Ocean power, including tidal energy, wave energy and marine current energy using the seas' powerful currents, are other potential sources of renewable energy. With the second highest tidal range in the world the UK could benefit significantly from tidal power. Hydrogen is also being developed as a major fuel and energy carrier.

OTHER BENEFITS Renewable energy sources produce few greenhouse gases and offer security of supply in a volatile world. Brazil's ethanol programme has also saved over US$50 billion in avoided fuel imports. In China renewables could reduce the environmental and health costs of fossil fuels, which have cost the country approximately 7% of its Gross Domestic Produce; this could rise to 13% by 2020. Renewable energy also offers economic growth and job creation with over 2.4 million new jobs created globally to date – more than all the jobs in the global oil and gas industries combined.

RENEWABLE ENERGY INVESTMENTS Renewable energy is growing – and growing fast. US$120 billion was invested in new renewable capacity in 2008. This is expected to rise to US$450 billion per annum by 2012 and US$600 billion per annum from 2020.

* 2006 figures

A renewable future

"

Just as automobiles followed horses, electric lights replaced gas lamps and more recently computers displaced typewriters, so can technological advances make today's smoke stacks and cars look primitive, inefficient and uneconomical.

Janet L. Sawin, Worldwatch Institute

"

Renewable energy sources currently supply just 13% of global energy. This is only expected to rise to between 14% and 23% by 2030, with fossil fuels still supplying about 80%. Yet the growth in renewable energy today and the high level of investment support a more positive outlook. Global investment has quadrupled in four years alone from US$30 billion in 2004 to US$120 billion in 2008.

United Nations' and other studies also affirm that renewables could supply over 50% of global energy by 2050 and all the world's energy long term. The future road we take will depend on funding, the growth of markets and consumer support, as well as the will of governments to effect rapid change and to distance themselves from fossil fuel interests for the sake of the Earth.

It's about priorities. Either climate change can be prioritised and changes made now to preserve our planet, or we will lose what we have. It's also about money and collaboration. Will governments provide funds to save our world? And can saving it be made our collective global priority?

Photo: Cameron Davidson/Corbis

Funds to meet the scale of the crisis

Current funds are small relative to the scale of the challenge.

Stern Review Report on the Economics of Climate Change (2006)
The Stern Review © Crown Copyright

World military spending is over US$1 trillion a year, the US war against terrorism has cost over US$850 billion, the global credit crisis mustered trillions of dollars and the UK's defence budget alone equates to approximately US$50 billion, about double the government funding for renewables worldwide. Funds are needed urgently to match the scale of the climate crisis and all the money in the world will not help if we leave it too late.

The Stern Review confirmed that the costs of taking action are less than the costs of not taking action. A rise of 2° to 3°C in the Earth's average temperature could cost up to 3% of global Gross Domestic Product (GDP) – US$3 trillion per annum by 2050. A 5°C to 6°C rise could cost 5% to 10% and by 2200 it could amount to 25% of world GDP. Yet the cost of taking strong action now is only 2% per annum of world GDP.

If nations commit 1% of public revenues, rising to 2% of GDP, low-carbon energy projects can be co-financed with the private sector. 1% of UK public revenues is about £6 billion, just one sixth of the £40 billion raised through fossil-fuel related taxes. Certainly revenues generated from taxing the problem should be applied to support the solutions. If G8 countries apply just 1% of GDP, US$350 billion could back low-carbon technologies every year.

Photo: REUTERS/Ceerwan Aziz

1% of the world's deserts covered in concentrated solar power mirrors could provide all global electricity.

Concentrated Solar Power Plant, Nevada Solar One, Boulder City, Nevada USA uses 760 parabolic troughs with more than 180,000 mirrors that concentrate the sun's rays.

Large-scale renewable energy systems

Much of the world's current energy stock and infrastructure needs to be replaced by 2020. Such replacements must be large-scale renewable energy systems, otherwise we will be locked into high emissions for decades to come.

Concentrated Solar Power (CSP) systems could provide all of the world's electricity, distributed through low-loss, high-voltage direct current power transmission lines. The side product of CSP is water – a lot of drinkable water – while the shade under the mirror systems could support agriculture.

Solar photovoltaic power (solar PV) technology should be scaled up so that large solar power plants could take over from fossil fuel power stations. Only an estimated US$28 billion of extra investment could make solar PV cost competitive – just a fifth of the cost of developing, assembling and running the International Space Station over ten years.

And with water (H_2O) covering 70% of our planet, the potential of hydrogen is vast. Carbon-neutral hydrogen produced using renewable energy sources could reduce emissions by almost 100%. Hydrogen is an energy carrier, so electricity from large wind farms could be converted to hydrogen and transported to where it is needed. Iceland is leading the way – by 2050 it aims to power its entire transport system, including boats, with renewably generated hydrogen fuel.

Photo: REUTERS/Marcelo Del Pozo

Nuclear power

The expansion of nuclear energy has also been proposed by some countries as a low-carbon energy source, but public support is still largely lacking. One European Union poll found only 12% of people supported nuclear, countered by 80% favouring renewable energy sources.

Nuclear power is capital intensive, requiring an initial investment of US$2 to US$3.5 billion for each reactor. Not only expensive and a poor investment, it is also beset with other problems – nuclear accidents such as at Chernobyl, low-level radiation, risk of terrorist attack, nuclear proliferation and costly and dangerous waste. Nuclear power plants also need water to cool the reactors and are generally sited next to major water sources like rivers or oceans. In a warming world of uncertain rainfall, higher temperatures and rising sea level, nuclear power could become an unreliable source of energy. This was shown in 2003 during the European heatwave, when the output of French nuclear power stations fell as water from the rivers was too hot to cool the reactors.

However, the main reason for avoiding a major expansion of nuclear energy is: why take the world and humanity from one potential crisis to another, when we have clean and sustainable alternatives?

Vika Chervinska, an eight-year-old Ukrainian girl suffering from cancer (April 2006). A UN report estimated the Chernobyl nuclear disaster would cause about 4,000 deaths. Greenpeace reported more than 90,000 people were likely to die due to radiation.

Photo: Oded Balilty/AP/Press Association Images

By 2030 China's emissions are expected to be double the total emissions from all industrialised countries including the US, the European Union and Japan.

Photo: Peter Parks/AFP/Getty Images

Developing countries' emissions

"If we can't influence China and India in their coming energy business we will be locked in and we will have to live with the consequences for half a century or more.

Dr Fatih Birol, Chief Economist, International Energy Agency (April 2007)"

In 2007 China overtook the US as the world's biggest emitter. By 2015 it will have installed as much energy generating plant as the European Union's 25 countries use today; 90% will be coal-fired, lasting 50 to 60 years. In 2006 China was building an estimated two new coal-fired power stations every week. Unless China and developing countries embrace a low-carbon future, the dangerous warming of the Earth cannot be prevented.

Such countries have the right to grow economically, to raise the poor's living standards and pollute equally on a per capita basis. A Chinese person produces less than one quarter of a US citizen's emissions, and someone in India just one eighteenth. One solution, 'Contraction and Convergence', proposes that each person on Earth be allocated an equal per capita share from a total global carbon budget. By a set target year, emissions would be reduced on the basis of fairness and equity. Article 4.5 of the United Nations Convention on Climate Change also commits developed countries to facilitate and finance the transfer of low-carbon technologies to developing countries and this must be enforced to the necessary level.

Photo: Joerg Boethling/Still Pictures

21% to 45% of global carbon dioxide emissions could be captured and stored by 2050.

Carbon Capture and Storage

If renewable energy sources supply over 50% of global energy by 2050 and if most of the emissions produced by fossil fuels are captured and stored, sizeable emissions reductions would be achieved. Up to 10 billion tonnes of carbon dioxide could be captured and stored by mid-century – more than the total global emissions for 2005.

Given China and India's expanding coal use, Carbon Capture and Storage (CCS) is crucial. Up to 80% to 90% of the carbon dioxide from power stations, industrial sites and refineries could be captured and then stored in old oil or gas wells, coal seams or saline reservoirs. A number of large-scale CCS projects already operate and 20 demonstration projects are in development. And Japan intends capturing one sixth of its emissions by as early as 2020. CCS has the potential to significantly reduce emissions but government support is needed to achieve the emissions reductions in time.

CCS has also been proposed as an emergency measure to avoid dangerous climate change. As biomass crops grow, they absorb carbon dioxide from the atmosphere. If, when sustainably produced biomass is burnt to produce energy, the emissions are captured and stored, carbon dioxide could actually be withdrawn from the atmosphere while producing energy.

Critically, existing coal-fired power plants should be phased out and only plants with CCS should be built today.

At Vattenfall's pilot CO_2 free power plant at 'Schwarze Pumpe' in Germany, CO_2 emissions are captured from coal and stored underground.

Photo: REUTERS/Hannibal Hanschke

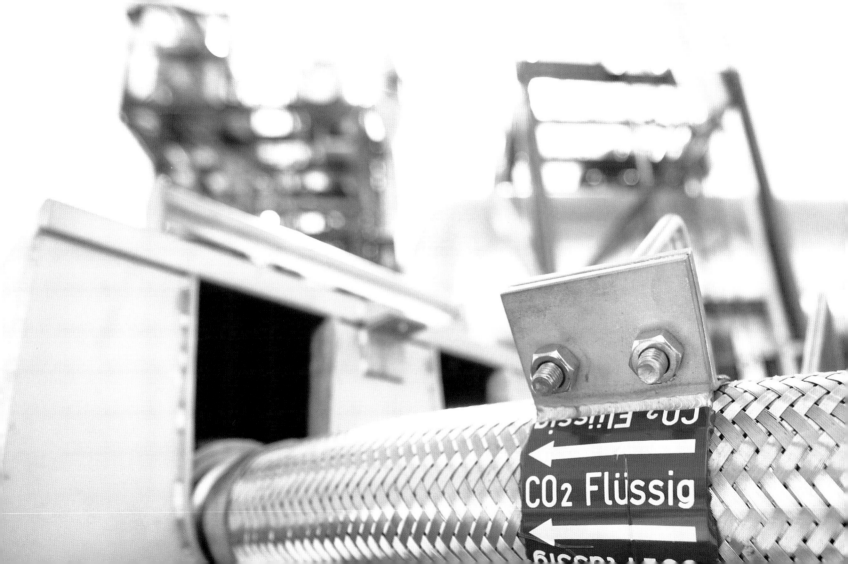

"Curbing deforestation is a highly cost-effective way of reducing greenhouse gas emissions and has the potential to offer significant reductions fairly quickly.

Stern Report on the Economics of Climate Change (2006)
The Stern Review © Crown Copyright "

Deforestation in Sarawak, Borneo, Malaysia.

Stop deforestation

8,000 years ago, half the Earth's land surface was covered in forests, compared to 30% today and every year a further 13 million hectares of forest are cut down. Not only do they no longer exist to absorb carbon dioxide from the atmosphere, but the carbon they store is released. Deforestation accounts for almost 20% of global carbon dioxide emissions – more than the global transport sector.

Preventing deforestation could however reduce emissions relatively quickly. It would also protect ecosystems. The main cause of deforestation is economic – forests are razed by multinational companies for export or the land converted for agriculture by big business or indigenous subsistence farmers. By giving incentives to landowners to protect their forests and by sharing advances in soil science that enable crops to grow on less productive ground, deforestation can be curbed. Debt-for-nature swaps are also effective. The US recently exchanged nearly US$30 million of Indonesian debt for protection of its forests.

Reducing Emissions from Deforestation and Forest Degradation in Developing Countries (REDD), is the proposal to include forest conservation in the new global treaty. 'REDD' or 'forest conservation units' could be traded on the global carbon market or direct compensation paid for forest protection. Compensation for the eight countries responsible for 70% of deforestation emissions is estimated at US$5 billion a year. Just US$0.5 billion per annum from each of the G8 countries, plus Brazil and China, could contribute to significant emissions reductions.

A worker harvests leaves from newly planted palm oil trees growing on the site of destroyed tropical rainforest in Kuala Cenaku, Riau Province, Indonesia (November 2007).

Photo: Dimas Ardian/Getty Images

Dr James Hansen, Director of NASA's Goddard Space Institute, has reported that the amount of carbon dioxide in Earth's atmosphere, which currently stands at about 387ppm, could be reduced to below 350ppm towards the end of this century.

This could be achieved by phasing out coal use except where the carbon dioxide is captured and stored and by adopting agricultural and forestry practices that extract greater amounts of carbon from the atmosphere and then store it in forests and soils.

Biochar produced from chicken waste and wood chips, has been turned into a valuable fertilizing substance that stores carbon. Wardensville, West Virginia (November 2008).

Improve agricultural and forestry practices

Global agriculture and related land-use change produce one third of greenhouse gas emissions through nitrogen-based fertilisers, cattle's digestive gases, biomass burning, rice production, manure and deforestation for agriculture, livestock and other uses. As food production increases to meet a growing population, such emissions could escalate. Alternatively, sustainable agricultural and forestry practices could help 'cool the planet'.

Carbon dioxide can be 'drawn down' out of the atmosphere and existing stores of carbon can also be sustained by preventing deforestation and loss of grasslands, by re-foresting and re-vegetating deforested and degraded land, by reducing tilling of soils, using high-carbon absorbing cropping systems, avoiding inorganic fertilisers and by embracing climate-friendly livestock production. Carbon sequestration in soils also has great potential. By burning crop, forestry and animal waste residues in a low-oxygen environment 'biochar' is produced, a dark carbon-rich charcoal. Added to soils, this could store carbon for hundreds of years or even for millennia, while at the same time improving soil fertility and boosting crop growth.

A new climate treaty must make improved agricultural and forestry practices a key tool for successful mitigation of climate change. Farmers and smallholders worldwide will also need financial incentives to adopt these practices, which will in turn improve agricultural productivity and aid sustainable development and adaptation to climate change.

A bio-digester in Argentina (December 2008). Methane emissions, captured from pig manure, are used to produce clean electricity or biogas that can be used in engines and boilers.

Photo: REUTERS/Enrique Marcarian:

Centralised power stations and national grids waste as much energy globally as is consumed by the whole world's transport sector.

Decentralised energy

By the time electricity reaches our home or workplace, two thirds has been wasted in producing and transporting it. Power generation should be decentralised by moving it away from large fossil fuel and nuclear power stations to produce power locally, using renewable energy and combined heat and power systems that produce electricity and provide heat, hot water and even meet cooling needs. In the UK, decentralised power could reduce overall energy emissions by at least 15% and by 2050 could supply over eight million homes, one third of today's estimated 25 million homes.

Woking Council in the UK has set the lead. By decentralising its energy supplies, over 14 years it reduced carbon dioxide emissions from council buildings by 80%, reduced energy consumption by 52% and reduced water consumption by 44%. It also saved one third of the money it had previously spent on energy and water. London is now looking to follow suit. By 2025, 25% of London's energy supply is to be met by decentralised energy and more than 50% by 2050.

Italy (October 2008): solar photovoltaic panels are installed on the Vatican roof. A total of 2,700 panels will produce enough electricity to heat or cool the 6000-seater Nervi-Paul VI Hall where papal audiences are held.

Reducing emissions in cities

Today more than 50% of people on Earth live in cities, which emit over 70% of global energy-related carbon dioxide emissions. As a result, carbon reduction in cities can make a big difference. Hundreds of cities worldwide are already taking action: addressing transport modes and energy efficiency in buildings, encouraging combined cooling, heat and power use and setting renewable energy and/or carbon dioxide reduction targets. Tokyo aims for renewables to supply 20% of its energy by 2020, Adelaide's goal is for zero emissions from buildings and transport and London's target is to reduce greenhouse gas emissions by 60% of 1990 levels by 2025. New York aims for a 7% reduction by 2012. Over 900 US mayors have now adopted New York's goal, representing 85 million people – over a quarter of the US population.

Urban planning that focuses on clean energy is also growing and in China, Dongtan, the first carbon-neutral city, is planned for half a million inhabitants. Collaborations to reduce emissions in cities are also now widespread, including the European Green Cities Network and the Large Cities Climate Leadership Group (C40), which involves 40 of the largest cities working together to tackle climate change. For smaller communities a movement called Transition Towns has set out steps that towns can follow in the transfer to a low-carbon society.

Governor Arnold Schwarzenegger speaks as the last of 1,727 solar panels are installed on the rooftop of the Staples Center sports complex in Los Angeles, California (October 2008).

As the largest economic sector, with US$4 trillion in annual premiums and US$55 trillion in managed assets, the insurance industry could play a key role in steering society towards a low-carbon path and so minimise insurance losses.

A house flattened by a US hurricane.

A role for the global insurance sector

2005 was the insurance sector's landmark year, with Hurricane Katrina's US$45 billion loss and US$80 billion losses worldwide. Economic losses were a staggering US$230 billion. Insurance payouts for weather-related disasters have risen 29-fold since the 1960s to an average US$22 billion per annum. And by 2040 or before, economic losses could hit US$1 trillion in a single year. With 80% of leading insurers worth less than US$0.5 billion, the industry is becoming increasingly vulnerable.

Some insurers are now reducing their exposure – in the US half a million policies in hurricane-prone Florida have been cancelled or not renewed, and New Orleans's recovery has been impeded by lack of insurance. In the UK after record flood losses in 2000, 2002 and 2007 combined with inadequate flood defence funding and homes being built on flood plains, some properties may now lose flood insurance cover.

Climate change could severely damage the insurance sector, but as the largest global industry it could steer society towards a low-carbon path. Insurers could lead with in-house carbon neutrality, stimulate customer take-up of low-carbon technologies by offering rewards of reduced premiums, and influence government policy and associated companies. Most importantly, with US$55 trillion of managed assets, it could support and so accelerate the expansion of low-carbon technologies.

The Swiss Re Tower, London, also called 'the Gherkin' but officially known as 30 St Mary Axe, was commissioned as the UK head office for Swiss Re, the global reinsurance company.

Adaptation

If all emissions were stopped today the Earth's climate would continue to change for over 50 years and the world would warm another 0.6°C. This means we are already committed to more severe climate impacts and must prepare now. Technology, climate-resilient buildings, flood- and drought-resistant crops, new planting techniques, sea defences and disaster preparedness can all reduce impacts and costs. And by integrating adaptation with disaster management and development policies, resources can be maximised and vulnerability reduced.

Developed nations have sufficient wealth to adapt, but developing countries have few resources. By 2100 at least 5% to 10% of Africa's Gross Domestic Product (GDP) will be needed to adapt to coastal flooding and sea-level rise alone. The cost of adapting to food and water scarcity, disease and disasters will be additional to this.

An estimated US$4 to US$86 billion of extra funding is needed annually to enable developing countries to adapt to climate change. US$86 billion is just 0.2% of developed country GDP. By 2007 however, only US$26 million had been spent through dedicated adaptation funds. Yet every US$1 invested in pre-disaster management can prevent losses of US$7, which means it is cheaper to act now than to wait. A Global Adaptation Fund is needed, unifying existing sources with increased funding procured through an aviation 'climate-levy' and 'climate taxes' to enable countries to adapt to climate change today.

Chinese villagers work on a concrete embankment to protect their county from the swollen Yangtze River in Wuhan (August 2002).

Photo: REUTERS/Claro Cortes

The technologies exist to power and protect our world.
What is needed is the will to create the change required.

People stroll pass the vast solar photovoltaic power system at the Universal Forum of Culutres, Barcelona, Spain
(April 2004)..

Photo: Alberto Estevez/epa/Corbis

The Will

Clean sources of energy are available now and would prevent the world warming to a dangerous degree. But action to date is wholly inadequate to the scale of the crisis we face. And the window of opportunity to protect the Earth is closing. What is needed is the will to make the changes required and to do so for our children and for the Earth. For our world is irreplaceable.

Germany has shown what can be achieved if a government wills to bring about change. In the early 1990s it had almost no renewables industry, but today it is a global leader in wind and solar photovoltaics. 14% of its electricity is now supplied by renewable energy and over 280,000 people work in the renewables sector, which has a turnover of over US$40 billion a year. To achieve this, the German government passed pioneering laws supporting renewables, provided low-interest loans and introduced feed-in tariffs and an Ecotax on fossil fuel sources. By 2050 it aims for at least 50% of its energy to come from renewables. Germany shows that if the will exists, change can happen quickly.

Photo: Ingo Wagner/epa/Corbis

"...the Swedish Government has set a new policy target: the creation of the conditions necessary to break Sweden's dependence on oil by 2020... If we prepare now, the transition to a sustainable energy system can be smooth and cost-efficient. If we wait until we are forced by circumstances, the transition may be costly and disruptive. No country can escape from this transition; to act sooner or act later are the only options.

Mona Sahlin, Minister for Sustainable Development for Sweden (May 2006)"

And that is the rub. We either transfer to a low-carbon economy now, or if we delay by just a decade, we change our world forever.

Change

With a climate-conscious US leader now at the helm, the hope is that global change can begin to accelerate. It needs a world leader and protector of the Earth to bring nations together to devise and implement a global emergency plan to save our world. But will such change occur in time – will world emissions peak by 2015 and then start to fall? Based on current performance and future projections, the answer for now is 'No'. And so we change the world forever.

Yet President Obama's election to office confirmed the power of the human spirit and the will of a populace for change. Climate change needs such a mass movement, a global mass movement – it needs the world's people to take action and to begin the mass transfer from fossil fuels to green sources of energy and in so doing break political deadlock, so that we may preserve the Earth.

People coming together have succeeded before: with India's home rule, against South Africa's apartheid, the civil rights movement in America. Only this time the scale is much greater and with a critical time imperative. It needs the world's people to do all we can individually and encourage and inspire collectively, to stimulate action at work and to do all we can at home. It needs those who are powerful to use their power to effect change, for those in business to showcase what can be achieved and for city leaders to turn their cities green. And governments should set their own houses in order and go carbon neutral. In all these places there are people, and it is the world's people who hold the Earth's future in their hands.

Democratic presidential candidate Senator Barrack Obama at the Xcel Center in St. Paul, Minnesota (June 2008) after winning the Democratic presidential nomination in a bid for change.

Photo: Craig Lassig/epa/Corbis

WHAT YOU CAN DO TO HELP

Transfer to green energy

Your help is urgently needed to limit the warming of our planet.

If there is just one thing you would do today for the sake of your children and their children in turn, for the sake of the Earth – please send just one email or make just one call to transfer to a green electricity tariff. This may cost you no more than your current supplier.

And this may, if you can persuade your friends and colleagues to do likewise, contribute to a chain of change, to spur the more rapid transfer from fossil fuels to a low-carbon society.

All new products need consumer support to ensure their market expansion.

Please sign up TODAY.

(See Appendix A on page 239)

Photo: REUTERS/Christine Muschi

"

We stand on the threshold of transforming the Earth. In coming decades we may either continue on our current path and cause the dangerous warming of our world or as a global community we can come together in a bid for change.

I hope people will transfer to a green electricity tariff today to support the expansion of renewable energy sources and also save energy at home and at work to reduce their carbon dioxide emissions. We have to take action now if we are to protect our planet and our children's future.

Susan Sarandon, Actress

"

Environmental activists and supporters at a demonstration at the United Nations Climate Change Conference in Nusa Dua on Bali (December 2007).

Five further actions you could take

1. CALL FOR ACTION
Call for a global plan of emergency action aiming to limit the world's average temperature rise to 1.5°C above its pre-industrial level. See Appendix B on this call for action, page 240.

2. SAVE ENERGY
Save energy at home, at work, at school, university or college, within business and government and at your local church or shop. See Appendix C on page 243 for what you can do.

3. SUPPORT COMPANIES AND POLITICAL LEADERS THAT ARE TAKING ACTION
Dupont has reduced its emissions by over 70% since 1990, HSBC is the world's first major bank to become carbon neutral; Marks & Spencer has committed to do so. So as you shop, back companies that are taking action and minimise your presence with the passive. Write a letter of support or condemnation to a company and please only vote for political leaders who aim to take real action.

4. PLANT A TREE – ONE OR MORE A YEAR
The United Nations Environment Programme (UNEP) originally aimed to plant one billion trees to fight climate change through re-forestation. Many billions have now been pledged and planted. Individually, or as a family, please plant one tree a year and join the UNEP billion trees campaign www.unep.org/billiontreecampaign

5. BEGIN A CHAIN OF CHANGE
And most importantly, please get five other people to do these things, to begin a chain of change.

Photo: Oz Hutchins

Tackling the threat of global warming may seem an impossible and hopeless task. But I am optimistic for three reasons. First I have experienced the commitment of the world scientific community. Secondly I believe the necessary technology is available. Thirdly I believe we have a God-given task of being good stewards of creation. For our fulfillment as humans we need not just economic goals but moral and spiritual ones. Reaching out for the goal of long-term care for our planet and its peoples could lead to nations working together more effectively and closely than is possible with other goals.

Sir John Houghton,
Former Co-Chairman of Working Group 1 of the Intergovernmental Panel on Climate Change

In December 1968, the Apollo 8 crew were the first humans to journey to the Earth's Moon and the first to photograph it from deep space. This famous picture of the Earth affirms the beauty of our planet and home.

Photo: Apollo 8/NASA

A CALL FOR CHANGE

From the deepest of oceans to the richest of rainforests' our world is beautiful and pulsates with life. But now through our use of oil, coal and gas to power our lives we are set to destroy our planet – decimate species, wipe out ecosystems, render vast areas drought-stricken and lifeless, exacerbate hunger, increase the suffering, melt the ice sheets and raise the level of the oceans. And why? Because we cannot change from old energy products to clean sources of energy in time. We have to do so – the stakes are too high. But it needs each one of us to make a stand and call for green energy. Nor must we must we allow politicians any doubt that they owe it to humanity and to the Earth to take adequate action before it is too late – or be held responsible for the Earth's demise.

We must act now. For once our planet has heated up in the space of a few decades, we will not be able to turn the clock back. And so we commit all people on Earth and all who follow to a world in disarray, with increasing disasters and suffering. There is still a small window of time to prevent this, but only a very few years. Global emissions of carbon dioxide must peak by 2015 or before if dangerous climate change is to be prevented.

Please think of your children and the world they will inherit.
Please think of the Earth and what is at stake.
Please join hands and take action.

Our world is precious, it supports us, it cannot be replaced.

Design: Johan Steneros Adstrakt
Photo: Spencer and Vibeke Montero vibeke@photito.com and the International Federation of Red Cross and Red Crescent Societies
Detailed credits are in the references.

"

If you had to look in your grandchildren's eye and say, yes I knew it was going to happen but I didn't bother to do anything about it – that would be a terrible thing.

In the past we didn't understand the effect of our actions – unknowingly we sowed the wind and now literally we are reaping the whirlwind – but we no longer have that excuse – now we do recognise the consequences of our behaviour – now surely we must act to reform it – individually, collectively, nationally and internationally before we doom future generations to catastrophe.

Sir David Attenborough

"

APPENDIX A: TAKE UP THE UK 20% INITIATIVE

Change to a green electricity tariff

Please begin to change the world to a renewably-powered society by making the change yourself to a green electricity tariff. And see if you can get your workplace, school, college, church or even your local shops to do the same. Simply follow the instructions below and sign up for green power *.

The challenge is to get 20% of UK households to change to a green tariff by 2012 when London stages the Olympics – the Green Olympics. That's five million UK homes showing that people want renewable energy.

*To switch your home to Ecotricity call 08000 302 302 or go online at www.ecotricity.co.uk/climate-change-book – see the inside back cover for more details.

APPENDIX B:
A GLOBAL CLIMATE CHANGE EMERGENCY PLAN AND BLUEPRINT

Please get five people to send a signed copy of this letter to the government.

I/We...

of (insert your address) ....

....

request that the Government pursues the implementation of a Global Climate Change Emergency Plan which has the overriding objective of aiming to limit the world's average temperature rise to 1.5°C above its pre-industrial level.

I/We ask that the government undertake to work with other nations and visionary parties in compiling a global blueprint for a new global treaty that will delineate how best to achieve this objective to protect our world, its peoples and species.

I/We ask that the following measures be included in such a blueprint:

1. That a global goal be established, aiming to limit the world's average temperature rise to 1.5°C above its pre-industrial level.

2. That all nations aim for global emissions to peak by 2015 or before and then decline, with a target of reducing global carbon dioxide emissions by 80% to 90% below 1990 levels by 2050. This should be towards a long-term target of near-zero emissions this century. Developed nations should aim for a near-zero carbon economy by 2050 to allow for growth in developing countries.

3. That energy-saving targets are set every decade to 2050 so that by this date projected energy consumption is reduced by 50% and that minimum energy efficiency standards are phased in for all new appliances, buildings, renovations, lightweight vehicles and in industry. Inefficient products should be phased out and energy efficiency labelling made mandatory on all relevant products, with educational programmes to inform the public.

4. That 50% to 60% of each nation's total primary energy be derived from renewable energy sources by 2050. Legally binding renewable energy targets should be set every decade to 2050, ensuring that a growing percentage of each nation's total primary energy is derived from renewable energy sources to meet this goal. Electricity and heat production targets should also be set.

5. That no new coal-fired power stations should be built without carbon capture and storage (CCS) and existing plants should be phased out. That an increasing percentage of homes and businesses utilise small-scale decentralised energy systems with the aim that by 2050, 40% to 50% of a nation's homes and businesses are so powered, supported by tax incentives and grants.

6. That improved forestry and agriculture practices be included as a key mitigation measure in the new global agreement to maximise the sequestration of carbon.

7. That 1% of the government's public expenditure, rising to 2% of GDP, be applied to a national Climate Change Fund to support renewable energy, energy efficiency and CCS in joint ventures with the private sector. The private sector should have an option to buy out the Fund's share in any capital project so that the monies are returned to the Fund. Anti-deforestation measures, improved forestry and agricultural practices and adaptation measures should also be supported by the Fund.

8. That a 'climate tax' be imposed on all fossil fuels with the revenues applied to the Climate Change Fund. That all fossil fuel subsidies be transferred to renewable energy technologies by 2015 and all fossil fuels include in their cost the external damage they cause to the environment and health.

9. That a high 'climate levy' be imposed on international aviation fuel or an 'en route' emissions charge be introduced on all international flights. All revenues should be applied solely to a Global Adaptation Fund that unifies existing funds and is used to support adaptation measures in developing countries, with the overall aim of creating an annual US$86 billion fund supported by other sources.

10. That a Climate Change Minister and Department be put in place in all countries just as in the UK, to ensure the rapid expansion of renewable energy and the implementation of all climate change measures. It should ensure that all government sectors are working in a coordinated manner towards the climate change objectives and that over a five-year term all government departments and buildings become carbon-neutral.

Signed: ....

APPENDIX C: SAVING ENERGY

We all have a responsibility to each other, and especially to those who will suffer most from climate change, to think carefully about how we live our lives, and where possible to leave a smaller carbon footprint.
Thandie Newton, Actress

How you can help by saving energy at home and at work.

• Use energy-efficient light bulbs – they use up to 80% less energy than conventional bulbs and last up to ten times longer.

• Turn off lights and appliances when not using them in homes and offices – don't leave products on standby – turning lights off can reduce lighting costs by 15%.

• Only use as much as you need – don't burn away energy. If the washing machine is only half full use the half load/economy programme. Don't put more water in your kettle than your cup needs and put a lid on saucepans.

• Turn the thermostat down just 1°C and wear a jumper – this could save up to 10% on your heating bills.

• Leave space round radiators and don't waste energy heating unused rooms.

• Buy energy-efficient appliances – the most efficient use about 50% to 80% less energy than older models.

• Insulate your home – don't pay for energy that is heating the sky – roofs, walls and even the floor can be insulated and windows double glazed.

- Think about your means of transport – please think of your children's future and get rid of your gas-guzzling car. Change to a more efficient car with low CO_2 emissions if you can, try to use the car less, walk, cycle or use public transport, car share where possible and think twice about going so far for your holiday. Please avoid flying as much as you can.

- Drive at 50mph instead of 70mph – this reduces fuel consumption by 30%. Keep tyres inflated at the correct pressure and turn off the air conditioning to further reduce fuel consumption.

- Buy less and buy locally where possible – three pairs of shoes is better than ten, second-hand can be as good as new. Be aware of the food miles in your purchase. Buy locally produced products where possible.

- Reduce, reuse and recycle.

- Green your money – put your money with an ethical bank, use a green mortgage, have a green pension and utilise ethical investments. You could also write to trustees of pension funds and fund managers, asking what their policy is to deal with climate change. This will put pressure on these sectors.

- If you build your own home – why not use a passive house design, which uses passive heat from bodies, the sun and appliances to warm the air? Houses today can be made highly energy efficient, which saves you money on your fuel bills.

Go to http://www.energysavingtrust.org.uk to see how you can save energy today.

And finally ...

This book is dedicated to Maya, my child, and the children of her generation. I asked her to write something about the world and what we were doing. Perhaps only the words of children will reach those politicians who refuse to effect the crucial change we need. Perhaps your child may want to write and draw something that you could send to your head of state. For it is their future, our children's future, that is at stake.

I want to stop the power stations working and clean them. I want to clear the smog I want to blow all the gas away I want the world to have enough water and enough food including the poor people in africa I want the sea to stop rising and get smaller to like it was when I was three. we want the world to stay beautyful because we like it just the way it is.

Maya 6 years old

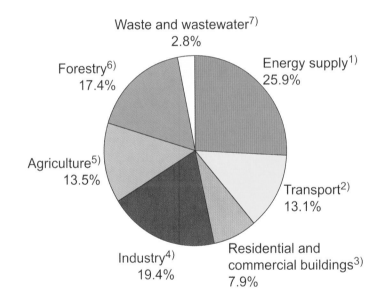

Greenhouse gas emissions by sector in 2004.

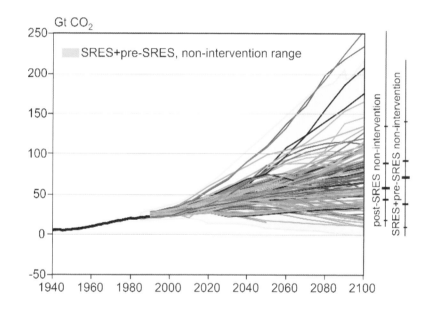

This graph shows the potential energy-related and industrial carbon dioxide (CO_2) emission scenarios to 2100. An emission of one billion tonnes of carbon (GtC) corresponds to 3.67 $GtCO_2$.

Ecotricity

Ecotricity began in 1995 with the very single-minded mission of changing the way electricity is made. Because conventional electricity is the UK's single biggest source of CO_2 and hence climate change, making the switch to green electricity is crucial if we are to avoid the worst effects of climate change – so clearly depicted in this book. This is our reason for being.

Ecotricity is an electricity company, but with a difference – we take the money our customers spend on their electricity bills and we re-spend it building new sources of clean power – windmills in fact. For every one pound our customers spend with us we spend another pound building windmills – we like to think of this as turning electricity bills into windmills.

We're recommended by Oxfam and the Soil Association and are the chosen supplier of nearly 40 thousand UK homes and businesses.

We can now all choose who supplies our electricity and where it comes from, so wherever you live in the country you can actually choose to have your home or business supplied by clean energy. The funny thing is switching takes about five minutes but it's the single biggest thing you can do to fight climate change.

Switching is easy. See the panel on the inside back cover of this book for details. When you switch to Ecotricity we'll even give you back the cover price of this book.

We hope you'll join us.

Photo: Ecotricity

REFERENCES

The Intergovernmental Panel of Climate Change (IPCC) was established by the United Nations Environment Programme and the World Meteorological Organization. Its role is to review and assess the work of thousands of scientists worldwide and to provide the world with a clear scientific view on the current state of climate change and its potential environmental and socio-economic consequences. The scientists who contribute to the IPCC's work do so on a voluntary basis.

This book draws on information contained in the most recent IPCC Fourth Assessment Report from 2007. Without the work of these very committed scientists we would not be in a position to safeguard our planet. Their work is to be much admired and respected and deserves the gratitude of the global community. Below are some of the key reports and information sources which have been utilised in compiling this book. Please go to www.ourworldfoundation.org.uk for the full references. With thanks and gratitude to the Intergovernmental Panel on Climate Change and all the scientists who work so very hard researching climate change.

IPCC, 2007: Climate Change 2007: The Physical Science Basis. Contribution of Working Group I to the Fourth Assessment Report of the Intergovernmental Panel on Climate Change [Solomon, S., D. Qin, M. Manning, Z. Chen, M. Marquis, K.B. Averyt, M. Tignor and H.L. Miller (eds.)]. Cambridge University Press, Cambridge, UK and New York, NY, USA, 996pp.
http://www.ipcc.ch/publications_and_data/publications_ipcc_fourth_assessment_report_wg1_report_the_physical_science_basis.htm
[Accessed 1.08.2009]

IPCC, 2007: Climate Change 2007: Impacts, Adaptation and Vulnerability. Contribution of Working Group II to the Fourth Assessment Report of the Intergovernmental Panel on Climate Change [M.L. Parry, O.F. Canziani, J.P. Palutikof, P.J. van der Linden and C.E. Hanson, (eds.)], Cambridge University Press, Cambridge, UK, 976pp.
http://www.ipcc.ch/publications_and_data/publications_ipcc_fourth_assessment_report_wg2_report_impacts_adaptation_and_vulnerability.htm
[Accessed 1.08.2009]

IPCC, 2007: Climate Change 2007: Mitigation. Contribution of Working Group III to the Fourth Assessment Report of the Inter- governmental Panel on Climate Change [B. Metz, O.R. Davidson, P.R. Bosch, R. Dave, L.A. Meyer (eds)], Cambridge University Press, Cambridge, United Kingdom and New York, NY, USA., XXX pp.
http://www.ipcc.ch/publications_and_data/publications_ipcc_fourth_assessment_report_wg3_report_mitigation_of_climate_change.htm
[Accessed 1.08.2009]

IPCC, 2007: Climate Change 2007: Synthesis Report. Contribution of Working Groups I, II and III to the Fourth Assessment Report of the Intergovernmental Panel on Climate Change [Core Writing Team, Pachauri, R.K and Reisinger, A. (eds.)]. IPCC, Geneva, Switzerland, 104 pp.
http://www.ipcc.ch/publications_and_data/publications_ipcc_fourth_assessment_report_synthesis_report.htm
[Accessed 1.08.2009]

IPCC, 2005: IPCC Special Report on Carbon Dioxide Capture and Storage. Prepared by Working Group III of the Intergovernmental Panel on Climate Change [Metz, B., O. Davidson, H. C. de Coninck, M. Loos, and L.A. Meyer (eds.)]. Cambridge University Press, Cambridge, United Kingdom and New York, NY, USA, 442 pp.
http://www.ipcc.ch/publications_and_data/publications_and_data_reports.htm#2   [Accessed 1.08.2009]

AOSIS 'Views on Shared Vision' Presented by Mr Philip Weech on behalf of The Alliance of Small Island Developing States AWG-LCA Shared Vision Workshop COP 14 Dec, 2 2008
http://copportal1.man.poznan.pl/Archive.aspx?EventID=26&Lang=english   [Accessed 1.08.2009]

Centre for Research on the Epidemiology of Disasters (CRED) www.cred.be
and EM-DAT: The OFDA/CRED International Disaster Database www.emdat.be
Université Catholique de Louvain, Brussels, Belgium   [Accessed 1.08.2009]

Cox P.M. et al (2006) 'Conditions for Sink-to-Source Transitions and Runaway Feedbacks from the Land Carbon Cycle' in H.J. Schellnhuber et al (eds) 'Avoiding Dangerous Climate Change' Cambridge University Press, February 2006, ISBN 9780521864718 DOI: 10.2277/0521864712

Dai, A. *et al* (2004) '*A global data set of Palmer Drought Severity Index for 1870-2002: Relationship with soil moisture and effects of surface warming*' J.Hydrometeoroly, 5, 1117-1130

Dlugolecki, A. (2006) UNEP Finance Initiative '*Adaptation and Vulnerability to Climate Change: The Role of the Finance Sector*'
sefi.unep.org/fileadmin/media/sefi/docs/briefings/CEO_Nov06.pdf   [Accessed 1.08.2009]

'*The Economics of Climate Change The Stern Review*' (2007) Nicholas Stern Cabinet Office – HM Treasury Cambridge University Press
http://www.hm-treasury.gov.uk/independent_reviews/stern_review_economics_climate_change/stern_review_report.cfm   [Accessed 1.08.2009]

Global Humanitarian Forum (2009) Human Impact Report: Climate Change '*The Anatomy of a Silent Crisis*'
http://www.ghfgeneva.org/OurWork/RaisingAwareness/HumanImpactReport/tabid/180/Default.aspx   [Accessed 1.08.2009]

Greenpeace (2006) '*Decentralising Power: An Energy Revolution for the 21st Century*' Greenpeace UK
www.greenpeace.org.uk/MultimediaFiles/Live/FullReport/7759.pdf   [Accessed 1.08.2009]

Hansen, J. *et al* (2008) 'Target atmospheric CO2: Where should humanity aim?' *Open Atmos. Sci. J.*, 2, 217-231,
DOI:10.2174187428230080201 0217   www.columbia.edu/~jeh1/2008/TargetCO2_20080407.pdf   [Accessed 1.08.2009]
Houghton, J. (2004) '*Global Warming –The Complete Briefing*' Third Edition, Cambridge University Press, Cambridge, UK

International Energy Agency (2007) '*Renewables in Global Energy Supply*' An IEA Fact Sheet, Jan 2007, OECD/IEA
International Energy Agency 9, rue de la Federation, 75739 Paris Cedex 15, France
www.iea.org/textbase/papers/2006/renewable_factsheet.pdf   [Accessed 1.08.2009]

International Energy Agency (2008) '*World Energy Outlook 2008*' OECD/IEA International Energy Agency 9, rue de la Federation, 75739 Paris Cedex 15, France www.worldenergyoutlook.org/   [Accessed 1.08.2009]

International Federation of Red Cross and Red Crescent Societies '*World Disasters Report*' 2008 P.O.Box 372 CH-1211 Geneva 19 Switzerland http://www.ifrc.org/Docs/pubs/disasters/wdr2008/WDR2008-full.pdf [Accessed 1.08.2009]

Jones, C. et al (2009) '*Committed terrestrial ecosystem changes due to climate change*' Nature Geoscience Letters Published online 28th June 2009 DOI:10.1038/NGEO555 http://www.nature.com/ngeo/journal/vaop/ncurrent/abs/ngeo555.html [Accessed 24.08.2009]

Martinot, E. and Sawin, J.L. (2009) '*Renewables Global Status Report 2009 Update*' REN 21 Renewable Energy Policy Network for the 21st Century http://www.martinot.info/ [Accessed 1.08.2009]

Meinshausen M. (2006) '*What does a 2C target mean for greenhouse gas concentrations? A brief analysis based on multi-gas emission pathways and several climate sensitivity uncertainty estimates*' in H. J. Schellnhuber et al (Eds.) '*Avoiding Dangerous Climate Change*' Cambridge University Press, February 2006, ISBN 9780521864718 DOI: 10.2277/0521864712

Meinshausen M. (2006) '*KyotoPlus-Papers <2°C Trajectories – a Brief Background Note*' Potsdam Institute for Climate Impact Research (PIK) Germany http://unfccc.int/resource/docs/2007/smsn/ngo/026d.pdf [Accessed 24.08.09]

Meteorological Office/Hadley Centre DEFRA Department of Environment Food and Rural Affairs '*Climate Change and the Greenhouse Effect – A Briefing from the Hadley Centre*' Dec 2005 http://www.metoffice.gov.uk/publications/brochures/ [Accessed 1.08.2009]

Munich Re, Geo Risks NatCatSERVICE http://www.munichre.com/en/ts/geo_risks/natcatservice/default.aspx [Accessed 1.08.2009]

Petit, J.R. et al (1999) '*Climate and Atmospheric History of the past 420,000 years from the Vostok Ice Core Antarctica*' Nature 399 pp 429-436 http://www.nature.com/nature/journal/v399/n6735/abs/399429a0.html [Accessed 1.08.2009]

Sawin, J.L. (2004) '*Mainstreaming Renewable Energy in the 21st Century*' Worldwatch Paper 169, State of the World Library http://www.worldwatch.org/node/821 [Accessed 1.08.2009]

Schellnhuber H.J., Cramer W., Nakicenovic N., Wigley T. and Yohe G. (eds) (2006) '*Avoiding Dangerous Climate Change*' Cambridge University Press, ISBN 9780521864718 DOI:10.2277/0521864712

Scherr, S.J. and Sthapit, S. (2009) '*Farming and Land Use to Cool the Planet*' in *State of the World 2009* pp 30-49 Worldwatch Institute, Earthscan London, UK

Thomas, C.D. *et al* (2004) 'Extinction risk from climate change' *Nature* 427, 145-148 DOI:10.1038/nature02121

UK Climate Projections UKCP09 http://ukclimateprojections.defra.gov.uk/content/view/2018/517/index.html   [Accessed 1.08.2009]

UNESCO (2009) '*The 3rd United Nations World Water Development Report: Water in a Changing World*' (WWDR-3) UNESCO Publishing and Earthscan, London, UK http://webworld.unesco.org/water/wwap/wwdr/wwdr3/index.shtml   [Accessed 1.08.2009]

United Nations Environment Programme GEO Year Book 2007 '*An Overview of Our Changing Environment*' http://www.unep.org/geo/yearbook/yb2007

United Nations Population (2009) '*World Population Prospects. The 2008 Revision*' http://esa.un.org/unpp/index.asp   [Accessed 1.08.2009]

Warren, R. (2006):'*Impacts of global climate change at different annual mean global temperature increases*', in H. Schellnhuber *et al* (eds) '*Avoiding Dangerous Climate Change*', Cambridge University Press, February 2006, ISBN 9780521864718 DOI: 10.2277/0521864712

Warren, R., Arnell, N., Nicholls, R., Levy, P., and Price. J. (2006): '*Understanding the regional impacts of climate change*', Research report prepared for *The Stern Review*, Tyndall Centre Working Paper 90, Norwich: Tyndall Centre, available from: http://www.tyndall.ac.uk/publications/working_papers/twp90.pdf   [Accessed 1.08.2009]

Acknowledgements

To the deeply committed scientists and individuals worldwide, whose work on the science, impacts and solutions to climate change is facilitating world action on global warming.

And with many many thanks to : Michael Tovey, Lyn Hemming and Alastair Sawday, Dale Vince, to my wonderful family and daughter Maya, Dr Andrew Dlugolecki, Dr Rachel Warren of the Tyndall Centre for Climate Change Research, Sir John Houghton, Dr R.K.Pachauri Chairman of the Intergovernmental Panel on Climate Change, Archbishop Desmond Tutu, Susan Sarandon, Thandie Newton, Sir David Attenborough, Dr Janet Sawin of the Worldwatch Institute, Dr Fatih Birol, Mona Sahlin, Tom Germain and Katherine Pate, David Oliver, Chris Wintle, Helen Johnson and Mark Neveu at Ecotricity, Mark Edlitz at Silly Goose Productions, Lavinia Browne, Jamie Harhay, Emily Rogath and Jillian Fowkes at ID PR, Dave Elliott of RENEW, Simon Bennett of Arnold and Porter, Chrissie Nyirenda, Margaret Martin, Tony Weyiouanna, Siniva Laupepa, Professsor Andrew Derocher, Monimala Sarker, Vibeke Montero, Amy Toensing, Jake Wall, Jan Grolinski, James Balog, Jason Hawkes, Schott Solar AG Power, Neil Crumpton, Cherry Alexander, Kim Lee, Fran Morales and Reuters Pictures, Nicholas Malherbe and all at Corbis, Matt Rowan and Press Association Photos, Kirsty Fuller and Getty Images, Susan Henry and National Geographic Stock, Sarah, Teresa and Julie-Anne Wilce at Still Pictures, Jorge Perez, Patrick Bosset, J.C. Chamois, Seija Tyrninokska and Aradhna Chadha at the International Federation of Red Cross and Red Crescent Societies, Mark Dowd at Topfoto, Bart at *Our Planet* magazine, Johan Steneros, Dominique Appasawmy and Justin Menschen at MMS, Dr Malte Meinshausen, Dr John Walsh of the International Arctic Research Center, Dr David Lawrence of the National Center for Atmospheric Research, Dr Eric Martinot, Allan Jones MBE, Kerry Alexander and Christopher Botten of the London Climate Change Agency, Giovanna Dunmail, Karen Lewis, Miguel Mendonca and Victoria Peat, David Hargitt, Barry Meehan, Philippe Hoyois and Debby Guha-Sapir from the Centre for Research on the Epidemiology of Disasters (CRED), Dr Eric Wolff, Dr David Vaughan, Dr John Turner and Peter Fretwell of the British Antarctic Survey, Angelika Wirtz and Petra Low of Munich Re Geo Risks Research NatCatSERVICE, Detlef van Vuuren of the Netherlands Environment Assessment Agency, Dr Eleonor Burke, Fiona Smith, Dr Chris Jones, Dr John Kennedy, Mireille Hartley, John Hammond and Mark Machin of the UK's Met Office and Hadley Centre, Paul van der Linden of the IPCC, Paul Dowling and Roberta Quadrelli of the International Energy Agency, Dr Dieter Luthi, Dr Jean Robert Petit, Dr Jean Marc Barnola, Dr A.Atiq Rahman and Sarder Shafiqul Alam of the Bangladesh Centre for Advanced Studies, Dr Saleemul Huq at the International Institute for Environment and Development, Dr Benito Mueller, Maria Beadle, Dr Pieter Tans, Dr Tom Delworth, Tom Smith and Dr Syd Levitus of NOAA, Carolin Arndt, Sophie Schlingemann, and Mary Jean Burer of WMO, Nic Williamson and Beth Williams of blah d blah, Chris Bird, Karen Crawshaw of BedZED, Johan Steneros of Adstrakt, Kakee Kaitu at the Tuvalu High Commission, Karsten Neuhoff, Paul Fogelmann, Paul Yiannouzis, Dave Williams, Erwin Northoff, TJ Blasing, Orville Huntingdon, Sheila Allen at *The Economist*, Dr William Nordhaus, Andrea Pocock of the United Nations, Daniel Crowe of the Renewable Energy Association Allan Doheny of the Canadian Ministry of Finance, Ann Doherty and Sisi of FoE International, Ed Gray, Lucy Martin of the Scott Polar Research Institute, Diarmid Campbell-Lendrum of WHO, Sam Leone of the Murray Darling Basin Commission, Lisa Elliston and John Hogan of ABARE, Clara Ion of the EU, Dr Mike Holland EMRC, Elizabeth Haylett of the Society of Authors, Andrea Meyer of the BMU. To anyone accidentally omitted please accept my apologies and sincere thanks for your help.